# MEDICINAL PLANTS OF INDIA

**[A Guide to Ayurvedic & Ethnomedicinal uses of plants]**

**With**

**Identity, Botany, Phytochemistry,**

**Ayurvedic properties, Clinical & Ethnomedicinal uses**

**Volume - I**

**DINESH JADHAV**

**SCIENTIFIC PUBLISHERS (INDIA)**

P.O. BOX 91 JODHPUR

*Published by:*
Pawan Kumar
Scientific Publishers (India)
5-A, New Pali Road, P.O. Box 91
Jodhpur - 342 001 (Raj.)
Tel.: +92-291-2433323, 2624154
E-mail: info@scientificpub.com
www.scientificpub.com



ISBN: 978-81-7233-545-8 (Set)
978-81-7233-546-5 (Vol. 1)
978-81-7233-547-2 (Vol. 2)

*Typed by:*
Narendra Parihar (Johny)
Parihar Solutions, Ratlam

Lasertype Set: Rajesh Ojha
Printed in India

*Dedicated to*
*The Memory of My Grandmother*
Late Mrs. **Kanchan Devi Jadhav**

# PREFACE

Medicinal plants belong to the earliest known health care products that have been used by the mankind. In India, the earliest reference to the medicinal value of plants appears in *Rigveda*, in which a brief reference to the healing property of plants has been made. However, it is in the *Ayurveda* that definite properties and uses of drugs and drug yielding plants have been discussed. In fact, Ayurvedic medicine had its origin from the works of *Ayurveda*. The later works of *Charak* and *Susruta*, namely *Charak-Samhita* and *Susruta Samhita* added invaluable knowledge to the science of medicinal plants.

Most of the drug plants are found in the tropics growing in wild condition and are mainly used by herb doctors and Ayurvedic vaidyas, who refer to them as *jari-butis*. In the modern age with the fast emergence of western modern medicine and synthetic drugs, the importance of plants and herbs started declining as more and more attention was paid to the development of synthetic drugs. However, with these synthetic drug came the problems of side effects and other clinical complications; the reason in the increasing awareness about the limitations of thc synthetic drugs. Because of natural herbal drugs are safe and effective; now herbal medicine and natural products are in big demand all over thc World.

This book is an effort to put together the vast variety of medicinal plants found in India. The plants selected in the present book are fairly widely used in India for millennia. This book well help in generating a global interest in *Ayurveda* and medicinal plants in India.

I hope that this compilation of most commonly used medicinal plants will be of use to Ayurvedic practitioners, students, Ethnobotanists and researchers as well as to medical and scientific communities.

**Dinesh Jadhav**

'Kanchan',
72 B, Rajiv (Kasturba) Nagar,
Ratlam 457 001 (M.P.)
E-mail: dinj2ad@yahoo.com

# ACKNOWLEDGEMENTS

First and foremost I express my deep sense of gratitude and indebtedness to Respected Dr. Shri. Anil K. Goel, Deputy Director, Botanic Garden, National Botanical Research institute (NBRI), Lucknow (Uttar Pradesh) for encouragement and valuable suggestions.

I offer my sincere thanks to all Respected Senior Professors of Govt. Arts and Science College, Ratlam (M.P.) for suggestions and kind cooperation in all respects. I owe my respect to all the non-teaching staff of Govt. Arts and Science College, Ratlam (M.P.) for valuable cooperation and encouragement.

It is also a pleasure to thank Mr. Manish Solanki, Library incharge, Central Library, Pt. Shivshakti Lal Sharma Ayurvedic Medical College and Hospital, Ratlam (M.P.) for allowing me to use the library.

Lastly, I take this opportunity to express my gratitude to Respected Shri. Pawan Kumar ji, Managing Editor, Scientific Publishers (India), Jodhpur, for accepting the Manuscript and also for guidance and helpful suggestions. Without his encouragement and support this book would not have been completed.

**Dinesh Jadhav**

# LIST OF PHOTOGRAPHS

# ABBREVIATIONS USED IN THE BOOK

| | | |
|---|---|---|
| A.ma. | : | Abhidhanamanjari |
| A. vi. | : | Ayurveda vigyan |
| Bha.pra. | : | Bhavprakash |
| Bri.ma. | : | Brind Madhav |
| Ca. ka. | : | Carakasamhita kalpasthanam |
| Ca.su. | : | Carakasamhita Sutrasthanam |
| Dha.ni. | : | Dhanvantarinighantu |
| Dra.gu.sam. | : | Dravyagunasamgraham |
| Dra.vi. | : | Dravyagunavijnan |
| Kai.ni. | : | Kaiyadevanighantu |
| Ni.ra. | : | Nighanturatnakaram |
| Ra.ni. | : | Rajanighantu |
| Sha. ni. | : | Saligramanighantu |
| Sa.ni.bhu. | : | Saligramanighantubhusanam |
| So.ni. | : | Sodhalanighantu |
| Syn. | : | Synonyms |
| Vern. | : | Vernacular name |

# KEY TO TRANSLITERATION

| Sanskrit | Transliteration key | Sanskrit | Transliteration key |
|---|---|---|---|
| अ | A, a | ट | ṭa |
| आ | Ā , ā | ठ | ṭha |
| इ | I, i | ड | ḍa |
| ई | Ī, ī | ढ | ḍha |
| उ | U, u | ण | ṇa |
| ऊ | Ū, ū | त | ta |
| ऋ | Ṛ, ṛ | थ | tha |
| – | Ṝ, ṝ | द | da |
| ए | E, e | ध | dha |
| – | Ē, ē | न | na |
| ऐ | Ai, ai | प | pa |
| ओ | O, o | फ | pha |
| – | Ō, ō | ब | ba |
| औ | Ou/ ou | भ | bha |
| अं | aṃ/ ama | म | ma |
| अः | aḥ | य | ya |
| क | ka | र | ra |
| ख | kha | ल | la |
| ग | ga | व | va |
| घ | gha | ष | ṣa |
| ङ. | naa | श | śa |
| च | ca | स | sa |
| छ | cha | ह | ha |
| ज | ja | ळ | ḷa |
| झ | jha | क्ष | kṣa |
| ञ | ña | ज्ञ | gna |

# CONTENTS

1. *Acacia arabica*

2. *Abutilon indicum*

3. *Achyranthes aspera*

4. *Adansonia digitata*

**Plate - 1**

5. *Adhatoda zeylanica*

6. *Aegle marmelos*

7. *Ageratum conezoldii*

8. *Ailanthus excelsa*

9. *Allium cepa*

10. *Aloe vera*

11. *Amaranthus spinosus*

12. *Asparagus racemosus*

**Plate - 2**

13. *Barleria prionitis*

14. *Calotropis gigantea*

15. *Calotropis procera*

16. *Cassia fistula*

**Plate - 3**

17. *Catharanthus roseus*

18. *Celosia argentina*

19. *Chenopodium album*

20. *Cissus quadrangularis*

21. *Clerodendrum phlomidis*

22. *Clitoria ternatea*

23. *Coccinia indica*

**Plate - 4**

25. *Datura stramonium*

24. *Cocculus hirsutus*

26. *Echinops echinatus*

27. *Eclipta prostrata*

28. *Euphorbia hirta*

29. *Ficus glomerata*

30. *Hibiscus rosa-sinensis*

31. *Holarrhena antidysenterica*

**Plate - 5**

32. *Hygrophila auriculata*

33. *Jatropha curcas*

34. *Lantana camara*

36. *Moringa oliefera*

37. *Mucuna pruriens*

35. *Lawsonia inermis*

38. *Nerium indicum*

39. *Nyctanthes arbor-tristis*

**Plate - 6**

40. *Ocimum basilicum*

41. *Ocimum sanctum*

47. *Ricinus communis*

42. *Phyllanthes emblica*

43. *Plumbago zelyniaca*

44. *Pongamia pinnata*

45. *Psidium guajava*

46. *Rauwolfia serpentina*

48. *Senna occidentales*

**Plate - 7**

49. *Senna tora*

50. *Tamarindus indica*

52. *Terminalia arjuna*

53. *Vernonia cineria*

51. *Tectona grandis*

54. *Vitex negundo*

**Plate - 8**

# INTRODUCTION

Man has depended on plants from time immemorial; plants have been a rich source for therapeutics. Medicinal plants belong to the earliest known health care products that have been used by the mankind. Over three-quarters of the World population rely on the use of traditional medicines for their primary health care needs. The medicinal plants are not only major components of the many formulations used in indigenous systems of medicine but also of a large number of drugs in Allopathy. In India, with one of the oldest ethnobotanical tradition in the World, medicinal plants are the backbone of all major systems of medicine. One of the earliest references to the Medicinal properties of plants is in the *Rigveda* (3500-1800 BC). *Ayurveda*, the ancient Indian system of medicine based on that medicinal property of plants. There are many more useful medicinal plants has been described under Ayurvedic medicinal system by several authors at that time i.e. *'Charakh'* and *'Susruta'* etc. Their Ayurvedic text manifests that an incredible 1600 of the 2000 old Ayurvedic formulations are plant based. After that there are many more books have been written under heading medicinal plants of India, but those books could not fulfil all the requirements of research scholar and readers. The present book is focused on medicinal plants of India. This book is differing from other books on medicinal plants because in this book each taxon contains ethno-medicinal uses with their Ayurvedic medicinal value. The ethno-medicinal value of the taxa is updated upto 2006. The sources of information included survey of published literature including primary, secondary and tertiary sources related to medicinal plants. The description on each plant species includes regional names, Distribution, Taxonomic description, Chemical constituents, Ayurvedic Medicinal properties, medicinal uses and Ethnomedicinal uses. Coloured photographs of some medicinal plants have been also included. This book is helpful for ethnobotanists, Ayurvedic medical practitioner, students and researchers as well as other reader's interested in the field of ethno-medicine. This book deals with 200 plants species of Angiosperms.

## ***ABELMOSCHUS MOSCHATUS*** (L.) Medic. (MALVACEAE)

***Syn.:*** *Hibiscus abelmoschus* L.

***Vern.:*** Kasturi Bhindi.

***Regional and Other Names :*** *Bengali* – Muskdana; *English* – Vegetable musk; *Gujarati* – Muskdana; *Hindi* - Maskdana; *Kannada* -Kasturi bende; *Malayalam* – Kattu kasturi; *Marathi* – Jangli bhendi; *Sanskrit* – Latakasturika; *Tamil* – Varattilaikasturi; *Telugu* - Kasturi benda.

***Distribution :*** Cultivated in the hotter parts of India.

***Taxonomic description :*** Strigose undershrubs, with ovate-cordate leaves. Flowers bright yellow with dark purple centre. Capsules acuminate, villous. Seeds reniform.

***Flowering & Fruiting :*** September – October.

***Medicinal parts :*** Seeds.

***Taste :*** Bitter.

***Chemical constituents :*** Seeds contains Farnesol, ambrettolide, sterols, terpenes and aliphatic compounds.

Seed oil contains 12, 13 – epoxyoleic acid, malvalic acid, sterculic acid & phospholipids.

***Medicinal properties :***

लता कस्तुरी तिक्ता स्वाद्वी वृष्या हिमा लघुः ।
चक्षुष्या छेदिनी श्लेष्मतृष्णावस्त्यास्यरोगहृत् ।। (भा.प्र.)

Latā kasturī tiktā svādvī vṛuṣyā himā laghuh
Cakṣhuṣyā chedinī śleṣmtṛiṣhṇāvaṣtyāṣyrōghṛiṭ. (Bhā. Pra.)

Seeds – Acrid, antispasmodic, aphrodisiac, aromatic, bitter, carminative, cooling, demulcent, digestive, diuretic, pectoral, stimulant, stomachic, tonic.

***Medicinal uses :*** Seeds - Dyspepsia, flatulence, gonorrhoea, hysteria, intestinal complaints, itching, leucoderma, nervous debility, snakebite, stomatitis, urinary discharges, vomiting.

***Ethno-Medicinal uses :*** Seeds - Insecticidal (Jain & Suri, 1980).

## *ABRUS PRECATORIUS* L. (PAPILIONACEAE)

***Syn.:*** *Glycine abrus* L.

***Vern.:*** Ratti.

***Regional and Other Names :*** *Bengali* – Kooch; *English* - Crab's eye; *Gujarati* – Gunja; *Hindi* – Ratti; *Kannada* – Ganji; *Malayalam* – Kunni; *Marathi* – Gunchi; *Sanskrit* – Aruna; *Tamil* – Gundumani; *Telugu* – Guravindha.

***Distribution :*** Throughout the tropics, often planted.

***Taxonomic description :*** Twining shrubs, with 10-20 pairs of leaflets and pink or white flowers having 9 stamens. Pods 3-5 seeded.

***Flowering & Fruiting :*** August-February.

***Medicinal parts :*** Leaves, Roots, Seeds.

***Taste :*** Bitter.

***Chemical constituents :*** Root contains Stigmasterol, Glycine, Glutamic acid, Aspartic acid, Tyrosin, Quercetin, Luteplin, Kaempferol, Taxalbumin; Alkaloid-abrin; Glucoside-abralin; Anthocyanin-abranin.

Seeds contain Lupeol acetate, Sitosterol, Sucrose, Gallic acid, Orietin, Isoorientin, Abrussic acid, Haemagglutinin.

Leaves contain Precol, Abrol, Abrasine, Precasine.

***Medicinal properties :***

गुंजाद्वयन्तु केश्यस्याद्वात पित्तज्वरापहम् ।
मुखशोषभ्रमश्वासतृष्णामदविनाशनम् ।
नेत्रभयहरं वृष्यं बल्यं कण्डू व्रणं हरेत् ।
कृमीन्द्र लुप्त कुष्ठानि रक्ता च धवलाऽपि।। (भा.प्र.)

Guṇjāḍvayāṇtū keśyaṣyāḍvāt piṭtajvarāpahaṃ
Mukhśōṣhbhṛamśvāsh ṭriṣhṇāmadvināśnuṃ
Neṭrabhayharaṃ vṛiṣhyaṃ baḷyaṃ kaṇdū vṛaṇuṃ hareṭ
Kṛumīṇdṛa lupta kuṣhthāni raktā ca dhavlāapi (Bhā. Pra.)

Root - Acrid, alexiteric, aphrodisiac, Bitter, diuretic, emetic.

Leaves - Acrid, alexiteric, aphrodisiac, Bitter, diuretic, emetic.

Seeds - Abortifacient, aphrodisiac, emetic, Purgative.

Fruits - Acrid, aphrodisiac, Bitter.

***Medicinal uses :*** Root - Asthma, biliousness, colic, fever, gonorrhoea, Jaundice, stomatitis, tuberculous glands.

Leaves - Asthma, cold, colic, cough, fever, gonorrhoea, headache, hoarseness, itching, jaundice, leucoderma, skin diseases, stomatitis, swelling, tuberculous glands.

Seeds - Diarrhoea, dysentery, leprosy, nervous disorders, paralysis, sciatica, ulcer.

***Ethno-Medicinal uses :*** Root - Abortifacient (Tarafder, 1983; Chevlan, 1998; Purkayastha & Nath, 2006); Snake bite (Chandra, 1995); Antiseptic, Bronchitis, Cold, Blood purifier (Goel *et al.,* 1984); Leucoderma, Promote Conception (Pal & Jain, 1989); Anthelmintic (Gaur *et al.,* 1980); Sterility (Purkayastha & Nath, 2006); Retention of urine (Kumar *et al.,* 2005); syphlis (Prakash & Singh, 2006.).

Leaves - Body pain (Sharma & Malhotra, 1984); Cough, Sore throat, Rheumatism (Bhalla *et al.,* 1982); Eye Complaint, Skin diseases (Janardhanan, 1963; Singh *et al.,* 1998); Diabetes (Nagaraju & Rao, 1989); Scabies (Pal & Jain, 1989); Leucoderma (Saxena, 1983); earache (Pereira, 2006); thorat infection (Mokat & Deokule, 2006); swelling (Bhalla *et al.,* 1982; Kumar *et al.,* 2006).

Seeds - Sterility (Jubilee Purkayastha & Nath, 2006). Abortifacient (Tarafder, 1983; Kapur, 1993); Antifertility (Bhalla *et al.,* 1982); Stomach complaint (Tarafder, 1983); Eczema (Purohit *et al.,* 1985; Upadhay *et al.,* 1998); Eye complaint (Bennet, 1985); Sciatica (Rajwar, 1983); Skin diseases (Janardhanan, 1963; Singh *et al.,* 1998); Fever (Srivastava & Rout, 1994); paralysis, sciatica (Rajendran & Rengamani, 2006).

Whole plants- Hepatitis (Dutta Choudhury & Shil, 2006).

## *ABUTILON INDICUM* (L.) Sweet (MALVACEAE)

***Syn.:*** *Abutilon asiaticum* (L.) Sweet.; *Sida guineensis* Schum.

***Vern.:*** Kanghi.

***Regional & Other Names:*** *Bengali* - Petari; *English* - Indian mallow; *Gujarati* - Dabali; *Hindi* - Kanghi; *Kannada* -Tutti; *Malayalam* –Tutti; *Marathi* - Mudra; *Sanskrit* – Atibala; *Tamil* – Nallatutti; *Telugu* - Adavabenda.

***Distribution :*** Throughout the tropics.

***Taxonomic description:*** Undershrubs, with ovate – suborbicular leaves and axillary, solitary flowers. Fruits with 15 – 25 mericarps.

***Flowering & Fruiting :*** September – January.

***Medicinal parts :*** Root bark, Leaves, Seeds.

***Taste :*** Sweet.

***Chemical constituents :*** Plant contains Gossypetin, Glucosides, Cyanidin, Rutinosides, essential oil.

Root contains Fatty acids, Aspargin, Rafinose.

Stem and Leaves contain Alkanol, Beta–Sitosterol, Vanillic acid, Caffeic acid, Fumaric acid, Fructose, Glucose, Galactose, Leucin, Histidine, Threonine, Serine, Glutamic acid and Aspartic acid.

***Medicinal properties :***

बलाचतुष्टयं शीतं मधुरं बलकान्तिकृत ।
स्निग्धं ग्राहि समीरास्त्रपित्तास्त्रक्षतनाशनम् ।। (भा.प्र.)

Balācatuṣtyam śītam madhrum balkāntikṛuṭ
Snigdhamgrāhi samīrāastrapittāstraksatnāśnam (Bhā.Pra.)

Root – Antipyretic.

Root bark - Astringent, diuretic.

Leaves - Demulcent, diuretic, emollient.

Seed - Aphrodisiac, demulcent, laxative.

***Medicinal uses :*** Root - Eye diseases, fever, leprosy, leucoderma, Piles, stone in bladder, strangury, uterine hemorrhagic discharges.

Root bark - Strangury, urinary complaints.

Leaves - Boils, bronchitis, diarrhoea, fever, gonorrhoea, gums, inflammation, lumbago, piles, toothache, ulcers, vaginal infections, wounds.

Seeds - Bronchitis, chronic chest trouble, cough, cystitis, gleet, gonorrhoea, piles.

***Ethno- Medicinal uses :*** Root - Paralysis (Jain & De, 1966); Cough (Singh & Pandey, 1980); Leprosy (Rajwar, 1983); Wound (Sudhakar & Chetty, 1998); Leucorrhoea (Jadhav, 2006); fever (Deokule, 2006).

Leaves - Boil, bronchitis, diarrhoea, ulcer (Bhalla *et al.*, 1982); asthma, spermatorrhoea (Kumar & Chauhan, 2005); Headache (Kumar *et al.*, 1980); Wound (Rajwar, 1983; Tripathi *et al.,* 1996); toothache (Aditya *et al.,* 2006); bleeding piles (Sharma *et al.,* 2006); gonorrhoea (Bhalla *et al.*, 1982; Sharma *et al.,* 2006); jaundice, piles (Rajendran & Rengamani, 2006).

Fruits - Stomach Complaint (Tarafder, 1983).

Seeds - Piles (Shah *et al.*, 1983; Mishra & Verma, 1995; Tripathi *et al.,* 1996); Bronchitis (Mishra & Verma, 1995; Tripathi *et al.,* 1996); Arthritis (Katewa & Sharma, 1998); Gonorrhoea (Mishra & Verma, 1995; Tripathi *et al.*, 1996).

## *ACACIA ARABICA* Auct. (MIMOSACEAE)

***Syn.:*** *A. nilotica* (L.) Willd. ex Delile ssp. *indica* (Benth.) Brenan.

***Vern.:*** Babul.

***Regional and Other Names :*** *Bengali* - Babla; *English* - Indian Gum Arabic Tree; *Gujarati* - Babalia; *Hindi* -Babul; *Kannada* - Gobli; *Malayalam* - Karuvelam; *Marathi* - Babul; *Sanskrit* – Ajabaksha, Babbula; *Tamil* - Karuveli; *Telugu* - Nallatumma.

***Distribution :*** Throughout the greater part of India.

***Taxonomic description :*** Trees, with long, white stipular spines. Pinnae 2-11 pairs and leaflets 7-25 pairs. Heads bright yellow, axillary, pedunculate, pods moniliform.

***Flowering & Fruiting :*** July – January.

***Medicinal parts :*** Stem, leaves, fruits, stem bark, gum, seeds.

***Taste :*** Acrid.

***Chemical constituents :*** Bark contains Beta-amyrin, Gallic acid, tannins, catechin, quercetin, leucocyanidin.

Fruit contains Gallic Acid, Catechin, Chlorogenic acid.

Gum contains Galactose, Arabic Acid, Calcium, Magnesium.

Seeds contains Ascorbic acid, Niacin, Thiamine, Amino acids.

***Medicinal properties :***

बब्बूलः कफनुद् ग्राही कुष्ठक्रिमिविषापहः। (भा.प्र.)

Babulaḥ Kaphnud grāhi kusthakrimiviśāpahaḥ. (Bha. Pra.)

Gum - Antidysenteric, astringent, styptic, tonic.

Stem bark - Acrid, alexipharmic, anthelmintic, aphrodisiac, astringent, demulcent, emetic.

Fruit - Acrid, astringent, cooling, expectorant, sweet.

Leaves - Antipyretic, astringent.

***Medicinal uses :*** Gum - Biliousness, bone fractures, diabetes, diarrhoea, dysentery, leprosy, sore throat, urinary discharges.

Stem bark - Biliousness, bronchitis, burning sensation, cough, diarrhoea, dysentery, leucoderma, piles, urinary discharges.

Fruit - Biliousness.

Leaves - bone fractures, bronchitis, eye diseases, gonorrhoea, piles.

***Ethno-Medicinal uses:*** Gum-Burns (Janardhana, 1963); Dysentery (Singh & Pandey, 1980; Basak, 1997); leucorrhoea (Kumar & Chauhan, 2005); Diarrhoea & Diabetes (Basak, 1997).

Stem Bark - Asthma & Bronchitis (Singh & Pandey, 1980; Sharma *et al.*, 2006); Diarrhoea, dysentery & diabetes (Bhalla *et al.*, 1982; Basak, 1997); Skin disease (Rajwar, 1983; Kumar *et al.*, 2006; Sharma *et al.*, 2006); bleeding piles (Kumar *et al.*, 2006; Sharma *et al.*, 2006); toothache (Deokule, 2006).

Fruit - Dysentery (Singh & Pandey, 1980); Impotency & Urino -genital disorders (Tripathi *et al.*, 1996).

Leaves - Eye complaint & throat infection (Singh & Pandey, 1980); Headache (Maheshwari & Singh, 1987); stomatitis (Kumar & Chauhan, 2005); Piles (Jadhav, 2006); Diarrhoea (Mohanty *et al.*, 1996); Gout (Jadeja *et al.*, 2006).

Flowers - Diarrhoea & dystentery (Singh & Pandey, 1980); Ear complaint (Saxena & Vyas, 1983).

Roots - Liver Complaint (Venugopal & Raghunathan, 1982).

Seeds- Spermatorrhoea (Prakash & Singh, 2006.).

## *ACACIA CATECHU* Willd. (MIMOSACEAE)

***Syn.:*** *Mimosa catechu* L. f.; *M. catechuoides* Roxb.

***Vern.:*** Kattha.

***Regional and Other Names :*** *Bengali* - Khayir; *English* - Cutch tree, catechu; *Gujarati* - Kherio; *Hindi* -Khair; *Kannada* - Kaggli; *Malayalam* - Kadaram; *Marathi* - Khair; *Sanskrit* – Balapatra, Khadira; *Tamil* -Kadiram; *Telugu* - Khadiramu.

***Distribution :*** Punjab, N. -W. Himalaya, central India, Bihar, Burma throughout the Konkan.

***Taxonomic description :*** Trees; all part pubescent; prickles hooked. Leaves 2-pinnate, with glands on the petiole and between the pinnae; leaflets 10-30 pairs. Spikes pale yellow. Pods flat.

***Flowering & Fruiting :*** May – September.

***Medicinal parts :*** Stem bark, resinous extract of heart wood.

***Taste :*** Bitter, astringent.

***Chemical constituents :*** Inner wood contains Catechin & Catechutannic acid (Catechutannic acid contain about 50% Tannin).

***Medicinal properties :***

खदिरः शीतलो दन्त्यः कण्डुकासारूचि प्रणुत्।
तिक्तः कषायो मेदोघ्नः कृमिमेहज्वरव्रणान् ।। (भा.प्र.)

Khadiraḥ śītaḷo dantyaḥ kaṇḍūkāsārucipranuṭ
Tiktaḥ kaṣāyo medoghnaḥ kṛimimehajvaravraṇān.

**(Bha. Pra.)**

Stem bark - Acrid, anthelmintic, antidysenteric, antipyretic, astringent, bitter, cooling.

Resinous extract - Acrid, anthelmintic, aphrodisiac, astringent, bitter, hot, tonic.

***Medicinal uses :*** Stem bark - Anaemia, boils, bronchitis, cough, diarrhoea, diseases of the tooth, elephantiasis, erysipelas, heaviness, indigestion, inflammation, itching, leprosy, leucoderma, psoriasis, sore throat, skin diseases, ulcers, urinary discharges, worms.

Resinous extract - Cough, diarrhoea, leprosy, leucoderma, leucorrhoea, oedema, piles, throat diseases, toothache, ulcer, urinary diseases, vaginal discharges.

***Ethno-Medicinal uses :*** Stem bark - Abortifacient (Tarafder, 1983); diarrhoea (Sharma *et al.,* 1979); leucorrhoea (Tarafder, 1985); gonorrhoea (Kapur, 1993); stomachache (Jadeja *et al.,* 2006; Kumar *et al.,* 2006).

Resin - Diarrhoea (Borthakur *et al.,* 1996); diabetes (Ghosh, 2006).

Wood - Skin diseases (Upadhay *et al.,* 1998).

Leaves - Dysentery (Maheshwari & Singh, 1987); skin diseases (Gupta *et al.,* 1999); pyorrhoea (Aditya *et al.,* 2006).

Flower - Gonorrhoea (Singh & Pandey, 1980).

Root - Rheumatism (Tarafder & Rai Chaudhuri, 1981); toothache (Mishra & Billore, 1983).

## *ACALYPHA INDICA* L. (EUPHORBIACEAE)

***Syn.:*** *Acalypha spicata; A. ciliata.*

***Vern.:*** Kuppi.

***Regional and Other Names :*** *Bengali* - Muktajhuri; *English* - Indian acalypha; *Gujarati* - Vanchhikan; *Hindi* - Kuppi; *Kannada* - Kuppameni; *Malayalam* - Kuppigidda; *Marathi* - Arittamunjariye; *Sanskrit* – Kuppameni, Rudra; *Tamil* - Kuppachettu; *Telugu* – Kuppinata.

***Distribution :*** Throughout the hotter parts of India.

***Taxonomic description :*** Erect herbs, with ovate or ovate – rhomboid leaves. Flowers greenish, in androgynous spikes; bracts concealing the capsules, Shortly dentate, capsules hispid.

***Flowering & Fruiting :*** June – September.

***Medicinal parts :*** Whole plant.

***Taste :*** Bitter.

***Chemical constituents :*** Plant contains Kaempferol, Acalyphamide, sterols, cyanogenic glyceride.

***Medicinal properties :***

मुक्ता वर्चास्तथा रूद्रा वान्तिकुच्च विरेचनी ।
कासश्वासगरघ्नी च ज्वरहृत् कफवातनुत् ।। (द्रव्यगुण)

Muktā varcāstathā rudrā vantikracca virechinī
Kāsasvasāgaraghni ca jvarahrt kaphavātanut. (Dra. Vi.)

Whole plant - Anthelmintic, carminative, diuretic, emetic, expectorant, laxative.

***Medicinal uses :*** Whole Plant - Asthma, arthritis, bronchitis, burns, earache, pneumonia, ringworm, scabies, skin diseases, snake-bite, toothache.

***Ethno- Medicinal uses :*** Whole plant - Cough & asthma (Nagendra prasad & Abraham, 1984; Chatterjee & Bhatterjee, 2006); pneumonia & rheumatism (Bhalla *et al.*, 1982; Sharma *et al.,* 2006); expectorant (Deokule, 2006); bronchitis (Bhalla *et al.*, 1982; Chatterjee & Bhatterjee, 2006; Sharma *et al.,* 2006).

Leaves - Scabies (Bhalla *et al.*, 1982; Arinathan *et al.,* 2006); Earache & snake-bite (Mishra & Sahu, 1984); Skin diseases (Shah, 1984; Tripathi *et al.*, 1996; Kiruba *et al.*, 2006; Ganesan *et al.*, 2006); ulcer (Jha *et al.*, 1997); wounds (Upadhaya & Jaswant Singh, 2005); itching (Rajendran & Rengamani, 2006).

Root - Snake-bite (Anand, 1996).

Flower- Gout (Jadeja *et al.,* 2006).

## *ACHYRANTHES ASPERA* L. (AMARANTHACEAE)

***Vern.:*** Apamarg.

***Regional and Other Names :*** *Bengali* - Apang; *English* -The prickly chaff- flower; *Gujarati* - Agheda; *Hindi* –Latjira; *Kannada* - Uttarane; *Malayalam* - Katalati; *Marathi* - Aghara; *Sanskrit* - Apamarg; *Tamil* - Nayurivi; *Telugu* – Apamargamu.

***Distribution :*** Throughout India.

***Taxonomic description :*** Erect hairy herbs, with oval–round pubescent- villous leaves. Flowers rather plump in appearance, deflexed, spinescent, in lax spikes.

***Flowering & Fruiting :*** August- February.

***Medicinal parts :*** Whole plant, Leaves, Roots, Seeds.

***Taste :*** Bitter & acrid.

***Chemical constituents :*** Seed contains Saponin A & B, Ecdy steron, oligosaccharide.

Root contains Oleonilic acid, maslinic acid.

Unripe fruit contains Saponin C & D.

Plant contains Nitrogenous compound- Achranthine & alkaloids.

***Medicinal properties :***

अपामार्गः सरस्तीक्ष्णो दीपनस्तिक्तकः कटुः ।
पाचनो रोचनश्छर्दिकफमेदोऽनिलापहः ।।
निहन्ति हृद्रुजाध्मानर्शः कण्डुशूलोदरापचीः । (भा.प्र.)

Apāmārgḥ Sarastīkṣhṇo dīpnastiktakaḥ kaṭuḥ.
Pāchno rocanschardikapmodoSnilāpahaḥ
Nihanti hṛdrujāghmārshaḥ kaṇḍuśulodrāpachiḥ. (Bha. Pra.)

***Medicinal uses :*** Root Abdominal pain, blood diseases, cancer, colic, deafness, disease of the tooth, heart diseases, itching, leucoderma, liver disorders, stone in bladder, tumor, vomiting, wound.

Leaves - Gonorrhoea, leprosy.

Seeds - Cold, colic, cough, hydrophobia, leprosy, scabies, toothache.

***Ethno- Medicinal uses :*** Roots - Whooping cough (Jain, 1984; Sharma *et al.,* 2006); Fever (Jadhav, 2006); Malaria (Singh, 1997); Abortifacient (Jain, 1965); Antifertility (Tarafder, 1983); Snake-bite (Nagendra Prasad & Abraham, 1984; Sharma *et al.,* 2006); Child birth (Tarafder, 1984); Pneumonia (Bennet, 1985); Dysentery (Saxena & Vyas, 1983); wounds (Jain & De, 1966); Cough (Singh & Pandey, 1986; Sharma *et al.,* 2006); Toothache (Jain, 1965; Ganesan *et al.*, 2006; Sharma *et al.,* 2006); Asthma, Bronchitis, Cold, Scorpion bite (Goel *et al.*, 1984); Stomatitis (Malhotra & Moorthy, 1973); spermatorrhoea (Prakash & Singh, 2006.); abdominal tumour (Sharma *et al.,* 2006).

Seed - Cough & Asthma (Bhalla *et al.*, 1982); Headache (Mishra & Billore, 1983); Skin diseases (Rajwar, 1983); Pyorrhoea, toothache (Narain & Singh, 2006).

Stem - Snake-bite (Singh & Maheshwari, 1992); toothache (Aditya *et al.,* 2006); jaundice (Das *et al.,* 2006).

Inflorescence - Cold & cough (Lal & Yadav, 1983); dog bite (Upadhyay & Jaswant Singh, 2005).

Leaves - Colic (Shah *et al.*, 1983); Debility (Maheshwari *et al.*, 1980); Piles (Baruah & Sharma, 1984; Baruah *et al.,* 2006); Ear complaint (Maheshwari *et al.*, 1980); Scorpion bite (Audichya *et al.*, 1980); Skin diseases (Rajwar, 1983); Leucoderma (Gupta, 1981; Sharma *et al.,* 2006); Sore & Stomachic (Srivastava & Verma, 1981); eye diseases (Kiruba *et al.*, 2006); sexual debility, spermatorrhoea (Kumar & Chauhan, 2005); asthma (Baruah *et al.,* 2006); cold (Sharma *et al.,* 2006); cough (Shah *et al.*, 1983; Sharma *et al.,* 2006); jaundice (Das *et al.,* 2006); menstruation trouble (Das *et al.,* 2006).

Whole plant - Boils (Sudhakar & Rolla, 1985); child birth, Rheumatism, scabies, syphilis (Jain & Tarafder, 1970); Cough (Bedi, 1970); Liver complaint (Lal & Yadav, 1983); skin diseases (Kiruba *et al.*, 2006); Dog bite (Jain *et al.*, 1973; Kadavul & Jisha, 2006); Piles (Sudhakar & Rolla, 1985; Chelvan, 1998; Tripathi *et al.*, 1996; Basak, 1997); Colic (Das, 1995); Renal disorders (Sudhakar & Chetty, 1998); Dropsy (Karnich *et al.*, 1981; Chelvan, 1998; Tripathi *et al.*, 1996; Basak, 1997; Das, 1995); snake bite (Deokule, 2006); diarrhoea, dysentery (Kataria & Gauchan, 2006).

## *ADANSONIA DIGITATA* L. (BOMBACACEAE)

***Vern.:*** Gorakh-imli.

***Regional and Other Names :*** *Bengali* – Gorakamli; *English* - Monkey–bread tree; *Gujarati* - Gorak ambli; *Hindi* – Gorakhimli; *Kannada* – Magimavu; *Marathi* - Gorakh chinch; *Sanskrit* –Chitrala, Goraksi; *Tamil* – Anaipuliamaram; *Telugu* – Brahmaamlika.

***Distribution :*** Grown in many places in India. Indigenous in tropical Africa.

***Taxonomic description :*** Deciduous trees, With digitately 3 to 9-foliolate leaves and white, pendulous flowers. Fruits ellipsoid, woody.

***Flowering & Fruiting :*** March – May.

***Medicinal parts :*** Stem bark, fruits & leaves.

***Taste :*** Sweet / acidic.

***Chemical constituents :*** Pulp & fruit contains Free tarteric acid, potassium bitartrate.

Leaves contain Mucilage.

Stem bark contains Bitter crystalloid – Adansenin.

***Medicinal properties :***

गोरक्षी मधुराम्ला च शिशिरा दाहपित्तनुत् ।
विस्फोटवान्त्यतीसार विषमज्वरनाशिनी ।। (रा. नि.)

Gorakṣī madhurāmlā ca śiśirā dāhapittanut
Visphoṭavāntyati sāra viṣamajvaranāśinī. (Ra. Ni.)

Leaves - Astringent, febrifuge, sudorific, tonic.

Stem bark - Astringent, cooling, stomachic.

Fruit pulp - Aperient, appetiser, astringent, cooling, demulcent.

***Medicinal uses :*** Leaves - Burning sensation, earache, inflammation, ophthalmia, swelling.

Stem bark - Intermittent fever.

Fruit pulp - Bronchial asthma, diarrhoea, dysentery, haemoptysis, itching, urticaria.

***Ethno- Medicinal uses :*** Skin irritation & rheumatism (Shah & Gopal, 1992).

## *ADHATODA ZEYLANICA* Medic. (ACANTHACEAE)

***Syn.:*** *A. vasica* Nees.; *Adennanthera vasica*; *Justicia adhatoda* L.

***Vern.:*** Adusa.

***Regional and Other Names :*** *Bengali* – Basak; *English* – Vasak; *Gujarati* – Aduso; *Hindi* – Adusa; *Kannada* –Adusoge; *Malayalam* – Adalodakam; *Marathi* – Adulsi; *Sanskrit* – Amalaka, Vasaka, Vansa; *Tamil* – Adatodi; *Telugu* – Atarushamu.

***Distribution :*** Throughout India.

***Taxonomic description :*** Shrubs, with elliptic leaves. Flowers white or creamish- white in elongate spikes; corolla – tube mottled. Capsules 4 – seeded.

***Flowering & Fruiting :*** January – April.

***Medicinal parts :*** Whole plant – Leaves, flowers, root bark.

***Taste :*** Bitter – acrid, astringent.

***Chemical constituents :*** Leaves contains A bitter vasicine, Adhatodic acid, volatile oil, vasa, Resin, gum, sugar & yellow pigment.

Plant contains Alkaloids (Vasicoline, Adhatodine, Vasicolinine, Anisolin), Vasicine & Vasicinone.

***Medicinal properties :***

वसायां विधमानायामाशायां जीवितस्य च ।
रक्तपित्ती क्षयी कासी श्वासी किमवसीदति ।। (बृ.मा.)

Vāsāyām vidyamānāyāmāsāyām jīvitasya ca
Raktapitti ksayī kāsī svāsī kimavasīdati. (Bri. Ma.)

The plant is acrid, anti-inflammatory, anthelmintic, antileprotic, antiperiodic, antiseptic, antispasmodic bitter, bronchodilator, cooling, diuretic, expectorant, pungent, stimulant.

***Medicinal uses :*** Asthma, blood impurities, chronic bronchitis, cold, cough, diseases of the mouth, fever, gonorrhoea, haemorrhagic diseases, heart troubles, jaundice, leprosy, ophthalmia, respiratory disorders, rheumatism, skin diseases, tumors, urinary diseases, whooping cough.

***Ethno-medicinal uses :*** Root - Anthelmintic, antipyretic, antiseptic (Bhalla *et al.,* 1982); boils (Karnick *et al.,* 1981); eczema (Singh & Pandey, 1980); rheumatism, asthma (Tripathi *et al.,* 1996); fever (Singh, 1993); bronchitis (Singh, 1993; Tripathi *et al.*, 1996).

Seed - Boils (Singh & Pandey, 1980).

Leaves - Bronchitis (Bhalla *et al.,* 1982; Basak, 1997; Billore *et al.,* 1998); asthma (Nagendra Prasad & Abraham, 1984; Basak, 1997; Billore *et al.,* 1998; Purkayastha & Subhan, 2005); Malaria (Bhalla *et al.,* 1982; Deka *et al.,* 2006); Swelling (Janardhanan, 1963); Rheumatism (Rao & Jamir, 1982; Basak, 1997; Billore *et al.*, 1998); cough (Saxena & Vyas, 1983; Basak, 1997; Billore *et al.,* 1998; Purkayastha & Subhan, 2005; Pereira, 2006); cold (Basak, 1997; Billore *et al.,* 1998; Pereira, 2006; Kadavul & Jisha, 2006); diarrhoea, dysentery (Jadeja *et al.,* 2006); Hepatitis (Dutta Choudhury & Shil, 2006).

Whole plant - Cough, asthma, bronchitis, anaemia (Kumar *et al.,* 2006).

***AEGLE MARMELOS*** (L.) Corr. ex Roxb. (RUTACEAE)

***Vern.:*** Bilvapatra.

***Regional and Other Names :*** *Bengali* – Bela; *English* - Wood apple; *Gujarat* – Billy, Bil; *Hindi* – Bel; *Kannada* – Bilpatre; *Malayalam* – Vilvam; *Marathi* – Baela; *Sanskrit* –Adhararuha, Bilva, Sri; *Tamil* – Bilwa; *Telugu* – Maredu.

***Distribution :*** Wild in the sub-Himalayan tract, Central and S. India. Often planted all over India.

***Taxonomic description :*** Deciduous trees, armed with spines. Leaves 3- foliolate and flowers creamy – white in axillary panicles. Fruits globose with stony rind and mucilaginous pulp.

***Flowering & Fruiting :*** March – April.

***Medicinal parts :*** Whole plant, Roots, bark, leaves, fruits, flowers.

***Taste :*** Bitter – Astringent.

***Chemical constituents :*** Fruits contains Gum, pectins, reducing sugars, sugars, oil, and tannin.

Leaves & Root contains Reducing sugars, Tannins, Iron, Calcium, Magnesium, and Potassium.

***Medicinal properties :***

श्रीफलस्तुवरास्तिक्तो ग्राही रूक्षोऽग्नि पित्तकृत् ।
वतश्लेष्महरो बल्यो लघुरूष्णश्च पाचनः ।। (भा.प्र.)

Sriphalastuvarāstīktō grāhī ruksyōagni pittakṛut
Vatasḷesmaharō balyō laghurusnasca pācanaḥ. (Bha. Pra.)

Ripe fruit - Acrid, antipyretic, antiscorbutic, appetiser, aromatic, bitter, cooling, febrifuge, nutritive, stimulant, sweet, tonic.

Root - Sweet.

Leaves - Astringent, bitter, digestive, febrifuge, laxative.

***Medicinal uses :*** Ripe fruit - Diarrhoea, biliousness, dysentery, dyspepsia.

Root - Abdominal pain, colic, fever, dyspepsia, dysuria, nervous disorders, oedema, rheumatism, urinary trouble, vomiting.

Leaves - Deafness, inflammation, ophthalmia.

***Ethno- Medicinal uses :*** Ripe fruit - Constipation (Singh & Pandey, 1980); diarrhoea & dysentery (Shah, 1984; Jadeja *et al.,* 2006; Kataria & Gauchan, 2006); Stomach diseases (Jain & De, 1996); Fever (Singh, 1997; Baruah *et al.,* 2006); dysentery (Purkayastha & Subhan, 2006; Pereira, 2006; Deokule, 2006; Baruah *et al.,* 2006); asthma (Baruah *et al.,* 2006); diabetes (Mokat & Deokule, 2006; Ghosh, 2006); eye diseases (Rajendran & Rengamani, 2006); piles (Das *et al.,* 2006).

Root - Dogs bite (Saxena & Dutta, 1975); hysteria (Kumar *et al.,* 2006).

Leaves - Backache & vomiting (Bhattacharjee *et al.,* 1980); Jaundice (Goel *et al.*, 1984; Dutta Choudhury & Shil, 2006); wounds (Singh & Maheshwari, 1983); Eye complaint (Singh & Pandey, 1980); Diabetes (Shah,1984; Singh & Prakash, 1994; Purkayastha & Subhan, 2006; Ghosh, 2006); Anti-inflammation (Katewa & Arora, 1997); Sun stroke (Singh & Prakash, 1994); worm infection (Pereira, 2006); dysentery, fever, asthma (Baruah *et al.,* 2006); abdominal pain (Kadavul & Jisha, 2006); piles (Das *et al.,* 2006).

Bark - Snake bite (Singh & Prakash, 1994); Diarrhoea, Dysentery & Stomach disorders (Saxena & Dutta, 1975); intermittent fever (Kumar *et al.,* 2006).

## *AGERATUM CONYZOIDES* L. (ASTERACEAE)

***Vern.:*** Osdi.

***Regional and Other Names :*** *Bengali* – Uchunti; *English* - Goat weed; *Gujarati* – Ajgandha; *Hindi* - Sadoradi; *Kannada* – Nayitulasi; *Malayalam* – Appa; *Marathi* – Ghaneraosadi; *Sanskrit*- Visamusti; *Tamil* – Pumpillu.

***Distribution :*** Throughout India.

***Taxonomic description :*** An annual hairy herb; leaves ovate – cordate, serrate; heads white in terminal corymbs.

***Flowering & Fruiting :*** Round the year.

***Medicinal parts :*** Leaves.

***Chemical constituents :*** Essential oil, Saponins, Flavonoids, alkaloids, amino acids, sugars, Vitamin – A & B.

***Medicinal properties :***

विषमुष्टिः कटुस्तिक्तो दीपनः कफवातहृत् ।
कण्ठामयहरो रूच्यो रक्तपित्तार्तिदाहकृत् ।। (रा. नि.)

Visamustīḥ katustīktō dipanaḥ kaphavātaḥṛt
Kanthāmayaharō rucyō raktapittārttidāhakṛt. (Ra. Ni.)

Leaf - Styptic.

Root - Antilithic.

***Medicinal uses :*** Leaf - Eye diseases, Leprosy, tetanus, wound.

Root - Eye diseases.

***Ethno- Medicinal uses :***

Leaf - Antiseptic (Malhotra & Moorthy, 1973), leucorrhoea (Katewa & Arora, 1997), typhoid (Jain & Sahu, 1993); headache (Maheshwari *et al.*, 1980; Kapur, 1993); muscular Pain (Ramachandran & Nair, 1981); piles (Singh, 1988; Sharma *et al.,* 2006); ringworm (Upadhye *et al.*, 1986); scabies (Abraham, 1981); skin diseases (Bhalla *et al.*, 1982; Sharma *et al.,* 2006); wound (Goel *et al.*, 1984; Baruah *et al.,* 2006; Yesodharan & Sujana, 2006; Gupta, 2006; Das *et al.,* 2006); tetanus (Upadhyay & Jaswant Singh, 2005); Stomachache (Udayan *et al.,* 2006); cuts (Baruah *et al.,* 2006; Kumar *et al.,* 2006; Gupta, 2006; Sharma *et al.,* 2006; Das *et al.,* 2006)); leprosy (Kumar *et al.,* 2006; Sharma *et al.,* 2006); malaria (Deka *et al.,* 2006); boils, burns, fever (Sharma *et al.,* 2006); swelling (Das *et al.,* 2006); eye trouble (Das *et al.,* 2006).

Flower - Cancer (Karnick *et al.*, 1981).

Seed - Diarrhoea (Bhalla *et al.*, 1982).

Root - Gall bladder (Chhetri, 2005); epilepsy (Narain & Singh, 2006b); kidney stones (Chhetri, 2005; Sharma *et al.,* 2006).

## *AILANTHUS EXCELSA* Roxb. (SIMAROUBACEAE)

***Vern.:*** Mahanim.

***Regional and Other Names :*** *Bengali* – Maharukha; *English* -The tree of heaven; *Gujarati* – Motoaduso; *Hindi* – Maharuk; *Kannada* – Doddamara; *Malayalam* – Mattipongilyam; *Marathi* – Maharuk; *Sanskrit* – Aralu, Mahanimba; *Tamil* – Perumaran; *Telugu* – Peddamanu.

***Distribution :*** Indigenous in the Indian peninsula and often planted in various parts of India.

***Taxonomic description :*** Large trees, with pinnate leaves crowned at the end of branches. Flowers yellow, in terminal and axillary panicles. Fruit a samara, spindle shaped, 1-seeded.

***Flowering & Fruiting :*** December- April.

***Medicinal parts :*** Leaves, bark.

***Taste :*** Bitter – Astringent.

***Chemical constituents :*** Bark contains Glaucarubin, Glauca rubinone, Excelsin, Glaucarubol, Ailanthion, Ailantic acid, Malantahin, Tricontane, Hexatricontane.

Leaves contain Ailantic acid, Beta – sitosterol (vitexin).

***Medicinal properties :***

महानिम्बस्तु शिशिरः कषायः कटुतिक्तकः ।
अस्त्रदाहवलासघ्नो विषमज्वरनाशनः ।। (रा. नि.)

Mahānimbastu śiśiraḥ kasāyaḥ katuṭiktakaḥ
Asradāhavalāsaghnō visamjvaranāśanaḥ. (Ra. Ni.)

Stem bark - Anthelmintic, antidysenteric, aromatic, astringent, bitter, diarrhoea, febrifuge.

Leaves - Bitter, carminative, febrifuge, tonic.

***Medicinal uses :*** Stem bark - Asthma, bronchitis, chronic fever, diarrhoea, dysentery, dyspepsia, earache, leprosy, piles, skin diseases, vomiting, worm infestation.

***Ethno- Medicinal uses :*** Bark - Cough & Cold (Joshi, 1982; Sharma *et al.,* 2006); abortifacient (Tarafder & Chaudhuri, 1981); Bronchitis (Shah, & Gopal, 1986); Typhoid (Jadhav, 2006); asthama (Tripathi *et al.*, 1996); Rectal problem (Mohanty *et al.*, 1996); diarrhoea (Siddiqui & Hussain, 1991; Deokule, 2006); dysentery (Pandit *et al.*, 1996; Deokule, 2006; Jadeja *et al.*, 2006); fever (Sharma *et al.,* 2006).

Leaves - Malaria (Aminduddin *et al.,* 1993); Rheumatism (Shah *et al.*, 1981); diarrhoea, dysentery (Deokule, 2006).

## ***ALANGIUM SALVIFOLIUM*** (L.f.) Wang. (ALANGIACEAE)

***Syn.:*** *A. lamarckii* Thw.; *Grewia salvifolium* (L. f.) Suppl.

***Vern.:*** Ankol.

***Regional and Other Names :*** *Bengali* – Akarkanta; *English* - Sageleaved Alangium; *Gujerati* – Ankola; *Hindi* – Ankol; *Kannada* – Ankole; *Malayalam* – Irinjil; *Marathi* – Ankol; *Sanskrit* – Ankota; *Tamil* – Alangi; *Telugu* – Ankolamu.

***Distribution :*** Throughout the greater part of India.

***Taxonomic description :*** A small tree with spinescent branches and grey bark. Leaves simple, alternate, membranous, varying in size and shape. Flowers in axillary clusters, small, white, and fragrant. Fruits berry, globular, crowned with persistent calyx-limb, purplish-red to black.

***Flowering & Fruiting :*** February – May.

***Medicinal parts :*** Root, Root bark, Leaves, Fruits, seeds.

***Taste :*** Pungent.

***Chemical constituents :*** Flower contain Deoxytubulosine alkaloid.

Fruits contain Cephaeline, deoxytubulosine, alangside, sterol, N-methylcephaeline.

Seed contain Alamarckine, emetin, cephaeline, isocephaeline & other alkaloids.

Leaves contain Alkaloids, sterols & terpenoids.

Stem bark contain Tubulosine, cephaeline, psychotrine alkaloids.

***Medicinal properties :***

अंकोटकः कटुस्तीष्णः स्निग्धोष्णस्तुवरो लघुः ।
रेचनः कृमिशूलामशोफग्रहविषापहः ।
विसर्पकफपित्ता स्त्रमूषिकाहिविषापहाः ।। (भा. प्र.)

Anakotakaḥ katustīksnaḥ snigdhoṣnastuvaro laghuḥ
Recanaḥ kṛmisūlāmsophagrahavisāpahaḥ
Visarpakaphpittā stramuṣikāhiviṣāpahāḥ. (Bha. Pra.)

Root - Anthelmintic, laxative.

Root bark - Anthelmintic, astringent, bitter, emetic, pungent, purgative.

Fruit - Astringent, laxative, refrigerant.

Seed - Cooling, tonic.

***Medicinal uses :*** Root bark - Colic, biliousness, fever, skin diseases.

Leaves - Rheumatic pain.

Fruits - Haemorrhages.

Seed - Boils, Haemorrhages.

***Ethno- Medicinal uses :*** Root - Fever (Sudhakar & Rolla, 1985); skin diseases (Bedi, 1978); dyspepsia (Jadeja *et al.,* 2006); rheumatic pain (Basak, 2006).

Stem Bark - Blood pressure (Singh & Pandey, 1980); asthma (Singh & Prakash, 1994); skin diseases (Kumar, 1992); jaundice (Singh & Singh, 1992).

Fruits - Eye complaint (Jain, 1963); skin diseases (Sharma, & Malhotra, 1984); stomach complaint (Sebastian *et al.,* 1984).

Stem - Abortifacient (Goel *et al.*, 1984).

Seed - Fever (Singh *et al.,* 1997).

Leaves - Boils (Saxena & Dutta, 1975); eye complaint (Chaudhuri *et al.,* 1975); rheumatic pain (Basak, 2006).

## *ALBIZIA LEBBECK* (L.) Willd. (MIMOSACEAE)

***Syn.:*** *Mimosa lebbeck* L.; *Acacia speciosa* Willd.

***Vern.:*** Sirus.

***Regional and Other Names :*** *Bengali* – Shirish; *English* - East Indian walnut; *Gujerati* – Kaliosaras; *Hindi* –Siris; *Kannada* – Bagemara; *Malayalam* – Vega; *Marathi* – Shiris; *Sanskrit* –Barhapushpa, Sirisa; *Tamil* – Vagei; *Telugu* – Dirasana.

***Distribution :*** Throughout India.

***Taxonomic description :*** Deciduous trees, with 1-4 pairs of pinnae and 4-10 pairs of leaflets. Flowers pale-green, in umbellate heads. Pods flat, straw-coloured.

***Flowering & Fruiting :*** April – September.

***Medicinal parts :*** Stem bark, leaves, flowers, and seed.

***Taste :*** Bitter, astringent.

***Chemical constituents :*** Bark contains Tannins, saponin, Echinocytic acid, and ß- sitosterol.

Pods contain Saponins, Echinocytic acid, Olionilic acid, Albigenine & Albigenic acid, lebbekanin – C, Glucose & Rhamnose.

Seed contains Triterpenic saponin.

***Medicinal properties :***

शिरीषो मधुरोऽनुष्णास्तिक्तश्च तुवरो लघुः।
दोषशोष विसर्पघ्नो कासव्रणविषापहः ।। (भा.प्र.)

Śīriṣo madhurōanuṣnastiktaśca tuvarō laghuḥ
Doṣaśoṣa visarpaghno kāsavraṇviṣāpaḥ. (Bha. Pra.)

Stem bark - Acrid, alexiteric, anthelmintic, astringent, bitter, cooling.

Root - Astringent.

Flower - Aphrodisiac, emollient.

Seed - Aphrodisiac, astringent.

***Medicinal uses :*** Stem bark - Asthma, blood disorders, bronchitis, conjunctivitis, cough, diarrhoea, eczema, erysipelas, helmintic inflammation, itching, leprosy, leucoderma, malaria, migraine, Oedema, paralysis, piles, Rhinitis, sinusitis, skin diseases, snake bite, ulcer, urticaria, worm Infestation.

Root bark - Good for gums.

Leaves - Nightblindness, Ophthalmia.

Seed - Gonorrhoea.

***Ethno- Medicinal uses :*** Bark - Dysentery & diarrhoea (Bhalla *et al.,* 1982); eye sores (Sharma *et al.,* 1979); ulcer (Kumar *et al.,* 2006); joint pain (Rajendran & Rengamani, 2006).

Leaves - Nightblindness (Sudhakar & Rolla, 1985); piles & diarrhoea (Singh & Aswal, 1992).

Seeds - Gonorrhoea (Bhalla *et al.,* 1982); piles (Malhotra & Moorthy, 1973; Singh & Aswal, 1992); diarrhoea (Singh & Aswal, 1992); leprosy (Gopakumar *et al.,* 1991).

Root - Ulcer (Bhalla *et al.,* 1982).

Flower - Boils & swelling (Bhalla *et al.,* 1982; Kumar *et al.,* 2006).

## *ALLIUM CEPA* L. (LILIACEAE)

***Vern.:*** Pyaz.

***Regional and Other Names :*** *Bengali* – Pyanj; *English* – Onion; *Gujarati* – Dungari; *Hindi* – Piyaz; *Kannada* – Nirulli; *Malayalam* – Ulli; *Marathi* – Kanda; *Sanskrit* –Dirghapatra, Palandu; *Tamil* – Vengayam; *Telugu* – Nirulli.

***Distribution :*** Cultivated throughout India.

***Taxonomic description :*** An annual herb, with bulbous stem, fistular, linear leaves and white flowers in umbels.

***Medicinal parts :*** Leaves, Bulb..

***Taste :*** Pungent-Acrid.

***Chemical constituents :*** Bulb contains Glutamyl, Arginin, Cystein, Quercetin, Kaempferol, Methylcystein, Sulphoxide, Propylecystein, Saponins.

***Medicinal properties :***

पलाण्डुस्तु गुणोर्ड्गेता रसोनसहशो बुधैः ।
स्वादु पाके रसेनोष्णः कफक्रन्नातिपित्तलः ।।
हरते केवलं वातं बलवीर्य करो गुरूः । (भा.प्र.)

Palāṇdustu gunourgyetā rasonasadṛaśo budhaiḥ
Svādu pāke rasenoṣnaḥ kaphakṛmnātipittalaḥ
Harte kevalam vātam balavīrya karō guruḥ. (Bha. Pra.)

Aphrodisiac, diuretic, expectorant, stimulant.

***Medicinal uses :*** Anti-tumour, bronchitis, dysentery, gum swelling, Jaundice, night blindness, piles.

***Ethno- Medicinal uses :*** Bulb - Anthelmintic (Gaur *et al.,* 1980); asthma, eye complaint, bronchitis (Nagendra Prasad & Abraham, 1984); boils (Malhotra & Moorthy, 1973); ear complaint (Mahsehwari & Singh, 1984); itching (Purohit *et al.,* 1985); piles (Singh, 1988) ringworm (Purohit *et al.,* 1985); skin diseases (Shah, 1984); stomach diseases (Bhalla *et al.,* 1982); vomit (Rao & Jamir, 1982); wound (Bhall *et al.,* 1982); chicken pox, fever

(Ganeshan *et al.*, 2006); epilepsy (Deokule, 2006) constipation (Jadeja *et al.*, 2006); diabetes (Ghosh, 2006).

## *ALLIUM SATIVUM* L. (LILIACEAE)

***Vern.:*** Lahsun.

***Regional and Other Names :*** *Bengali* – Lasan; *English* – Garlic; *Gujarati* – Lasan; *Hindi* – Lasun; *Kannada* – Bellulli; *Malayalam* – Vellulli; *Marathi* – Lasoon; *Sanskrit* – Rasona; *Tamil* – Ullipondu; *Telugu* – Vellulli..

***Distribution :*** Cultivated throughout India.

***Taxonomic description :*** Bulbous herbs, each bulb with several fleshy bulblets. Leaves linear. Flowers white, in umbels.

***Medicinal parts :*** Leaves, Bulb.

***Taste :*** Acrid, sweet.

***Chemical constituents :*** Bulb contains A volatile oil which contains Allyl disulphide, Allyl-propyl disulphide and Poly sulphide; Allicin (antibacterial): Allicetion-I and Allicetion-II (Antibiotics).

***Medicinal properties :***

रसोनो बृहणो वृष्यः स्निग्धोष्णः पाचनः सरः ।
रसे पाके च कटुकस्तीक्ष्णो मधुरको मतः ।।
भग्नसंधानकृत्कण्ठयो गुरूः पित्तास्त्रवृद्धिदः ।
बलवर्ण करो मेधाहितो नेत्रयो रसायनः ।। (भा. प्र.)

Rasono bhṛhino vriṣyaḥ snigdhosnaḥ pācanaḥ saraḥ
Rase pāke ca kaṭukastikshno madhurako mataḥ
Bhagnasandhānkritkanthayo guruḥ pittāṣṭravriḍḍhidaḥ
Balvarnakarō meghāhiṭo netryo rasāyanaḥ. (Bha. Pra.)

Antibacterial, antiseptic, aphrodisiac, bone healing, digestive, diphoretic, diuretic, expectorant, pungent, rejuvinator, stimulant, stomachic, tonic.

***Medicinal uses :*** Asthma, bronchitis, chronic cough, ear trouble, epilepsy, flatulence, malaria fever, skin diseases, tuberculosis, ulcer.

***Ethno- Medicinal uses :*** Bulb - Bronchitis, dropsy, wound, ringworm, tomour (Bhalla *et al.*, 1982); fever (Nagendra Prasad & Abraham, 1984); ear complaint, respiratory diseases (Rai, 1985); eczema (Purohit *et al.*, 1985);

whooping cough (Lal & Yadav, 1983); diarrhoea (Jadeja *et al.,* 2006); epilepsy (Deokule, 2006); bodyache (Mokat & Deokule, 2006); diabetes (Ghosh, 2006); cold (Gupta, 2006); cough (Nagendra Prasad & Abraham, 1984; Gupta, 2006); Gout (Jadeja *et al.,* 2006).

Leaf - Piles (Bhalla *et al.*, 1982); rheumatism (Saxena & Vyas, 1983).

Whole plant – Diabetes, hypertension, gastritis, amoebiosis, cancer, rheumatoid arthritis (Kumar *et al.,* 2006).

## *ALOE VERA* L. (LILIACEAE)

***Syn.:*** *A. barbadensis* Mill.; *A. perfoliata* L. var. *veral*

***Vern.:*** Gwarpatha.

***Regional and Other Names :*** *Bengali* - Ghrita kumari; *English* - Indian aloe; *Gujarati* – Kadvikunvar; *Hindi* – Ghee - kanwar, gwarpatha; *Kannada* – Komarka; *Malayalam* –Kattavazha; *Marathi* – Korphad; *Sanskrit* – Kumari; *Tamil* - Bhottu – katrajhae; *Telugu* - China kata – banada.

***Distribution :*** Planted in Indian gardens.

***Taxonomic description :*** Perennial herbs, with fleshy mucilaginous leaves in rosette having marginal spines. Flowers yellow, born in racemes on long scapes.

***Flowering & Fruiting :*** December – February.

***Medicinal parts :*** Whole plant- dried juice & pulp of leaves.

***Taste :*** Bitter.

***Chemical constituents :*** Leaves contain Carbohydrates, Tannins, Proteins, Chromones (Abesin & Aloesone), Malic acid, Citric acid, Tartaric acid.

Jelly from leaves contain Acetylated glucomannans, Iron, Calcium, Potassium & Magnesium.

***Medicinal properties :***

कुमारी भेदिनी शीता तिक्ता नेत्रया रसायनी ।
मधुरा बृहणी बल्या वृष्या वातविषप्रणुत् ।
गुल्मप्लीहयकृद्वृद्धिकफज्वरहरी हरेत् ।
ग्रंथयाग्निदग्धविस्फोटपित्तरक्तत्वगामयान् ।। (भा.प्र.)

Kumārī bhedinī śitā tiktā netryā rasāyanī
Madhurā bṛnhinī baḷyā vṛṣyā vātvispṛaṇuṭ
Guḷmaplīhyakṛdvṛḍḍhikaphajvarharī hareṭ
Granthyāgnīdagdha visphoṭapittaraktatvagāmyāṇ. (Bha. Pra.)

Leaf pulp - Analgesic, anthelmintic, anti – inflammatory, carminative, digestive, purgative, stimulant, stomachic, tonic.

***Medicinal uses :*** Leaf pulp - Cold, colic, constipation, cough, fever, menstrual disorders, piles, rectal fissures, rheumatism.

***Ethno- Medicinal uses :*** Leaf pulp - Fever, liver complaint, constipation (Bhalla *et al.,* 1982); Skin diseases (Bhalla *et al.,* 1982; Singh *et al.,* 1998); Boils, gonorrhoea (Rain, 1985; Singh *et al.,* 1998); Rheumatism (Shah *et al.,* 1981); Wounds (Lakshamanan & Sankaranarayanan, 1998; Chatterjee & Bhatterjee, 2006); piles (Bhalla *et al.,* 1982; Deokule, 2006); dyspepsia (Jadeja *et al.,* 2006); pimples (Mokat & Deokule, 2006); boils, burns, swellings (Yesodharan & Sujana, 2006); urinary complaints (Arinathan *et al.,* 2006).

## *ALSTONIA SCHOLARIS* R. Br. (APOCYNACEAE)

***Syn.:*** *A. cuneata* Wall; *Echites scholaris* L.

***Vern.:*** Saptparn.

***Regional and Other Names :*** *Bengali* – Chattim; *English* – Devil tree; *Hindi*- Chatium; *Kannada* – Maddale; *Malayalam* – Aarilampala; *Marathi* – Satvin; *Sanskrit* – Saptaparna; *Tamil* –Elilaipillai; *Telugu* – Eedakulapala.

***Distribution :*** Throughout India.

***Taxonomic description :*** A large tree with whorled branches and leaves with bitter milky juice, white flowers.

***Flowering & Fruiting :*** December-June.

***Medicinal parts :*** Leaves, Bark, Wood.

***Taste :*** Bitter.

***Chemical constituents :*** Leaves contain Picralinol, Akuammigine, Tubotaiwine, Akuammicine.

Flowers contain Picrinine, Strictamine, Tetrahydro alstonic, Hexacosane, Lupeol, Beta-Amyrin, Palmitic acid, Urolic acid.

Stem bark contain Echitamine, Venoterpine-Glucoside.

***Medicinal properties :***

सप्तपर्णो व्रणश्लेष्मवातकुष्ठास्त्रजन्तुजित् ।
दीपनः श्वास गुल्मघ्नः स्निग्धोष्णस्तुवरः सरः ।। (भा.प्र.)

Saptparno vranślesmavātkusthāstrajantujit
dipanah śvāsagulmaghnah snigdhosnastuvarah sarah. **(Bha.Pra.)**

Antileprotic, astringent, blood purifier, liver tonic, sedative, stomachic.

***Medicinal uses :*** Dropsy, dysentery, gastro-intestinal troubles, malaria, ulcers.

***Ethno- Medicinal uses :*** Bark - Hypertension, heart disorders (Chhetri, 2005); Chest pain (Chaudhari *et al.*, 1982); Diarrhoea (Sharma *et al.*, 1979; Kataria & Gauchan, 2006); Snake-bite (Nagendra Prasad & Abraham, 1984); dysentery (Kataria & Gauchan, 2006); Eye complaint, malaria (Goel *et al.*, 1984); malaria (Baruah *et al.*, 2006; Deka *et al.*, 2006); toothache (Aditya *et al.*, 2006); stomach trouble (Das *et al.*, 2006).

Leaves- Boils, stomachache (Baruah *et al.*, 2006); headache (Das *et al.*, 2006).

Latex- Ulcer (Kadavul & Jisha, 2006); allergy (Das *et al.*, 2006).

## *ALTERNANTHERA SESSILIS* (L.) R. Br. ex. DC. (AMARANTHACEAE)

***Syn.:*** *A. triandra* Lam.; *A. denticulata* R. Br.; *A. nodiflora* R. Br.

***Vern.:*** Kanchari.

***Regional and Other Names :*** *Bengali* – Kanchari; *Gujarati* – Jalajambo; *Hindi* – Saranti; *Kannada* -Honagone soppu; *Malayalam* - Ponnanganni – Keeray; *Marathi* –Kanchari; *Sanskrit*- Matsyaksi; *Tamil* – Ponnangannikkirai; *Telugu* – Ponnagagantakura.

***Distribution :*** Throughout India.

***Taxonomic description :*** A prostrate or spreading, polymorphic herbs; stem grooved, greenish-purple, with white hairy appendages, much branched, often rooting at the nodes; leaves glabrous, elliptic, linear-oblong, cuneate at the base, acute; flowers in axillary, small, white or pink-tinged, spicate-heads; seeds minute.

***Flowering & Fruiting :*** Throughout the year.

***Medicinal parts :*** Whole plant.

Plant contains Iron, protein, hydrocarbons & sterols.

***Medicinal properties :***

मत्स्याक्षी ग्राहिनी शीतकुष्ठपित्तकफास्त्रजित् ।। (भा.प्र.)

Mąṭsyākṣi grāhini śhitkuṣthapittakaphāṣṭrjiṭ. (Bha. Pra.)

Dried plant - Abortifacient, febrifuge.

***Medicinal uses :*** Dried plant - Blood disorders, leprosy.

Root - Wound.

***Ethno- Medicinal uses :*** Plant - Boils & as an eye – wash (Tripathi *et al.,* 1996); stomach & gastric trouble (Subhangi & Patil, 2006); snake bite (Basak, 2006; Das *et al.,* 2006).

Leaf - Snake bite & scorpion bite (Sudhakar & Chetty, 1998; Das *et al.,* 2006); skin diseases (Upadhay *et al.,* 1998); gastric problem (Purkyastha & Subhan, 2006); improve eyesight (Udayan *et al.,* 2006)..

Root - Rickets & marasmus (Agarwal & Arya, 2005).

## *AMARANTHUS SPINOSUS* L. (AMARANTHACEAE)

***Vern.:*** Chauli.

***Regional and Other Names :*** *Bengali* – Kantanotya; *English* - Prickly amaranth; *Gujarati* – Kantalodambho; *Hindi* - Kantewali chauli; *Kannada* – Mulladantu; *Malayalam*– Mullacherucheera; *Marathi* – Kantemath; *Sanskrit* – Alpamarisha, Tanduliya; *Tamil* – Mullukeeraj; *Telugu* –Nalladoggali.

***Distribution :*** Throughout India.

***Taxonomic description :*** Monoecious herb, with ovate-rhomboid or elliptic leaves having axillary spines. Flowers in axillary clusters, passing to spikes. Fruits ovoid.

***Flowering & Fruiting :*** August– December.

***Medicinal parts :*** Root and leaves.

***Taste :*** Sweet.

***Chemical constituents :*** Plant contains Carbohydrates, Proteins, Minerals, Iron, Vitamin – A and Calcium (Leaves).

***Medicinal properties :***

तण्डुलीयो लघुः शीतो रूक्षः पित्तकफास्त्रजित् ।
सृष्टमूत्रमलो रूच्यो दीपनो विषहारकः ।। (भा.प्र.)

Taṇdulīyo laghuḥ śito rūkṣhaḥ pittkaphāṣṭrajiṭ
Sṛṣṭmūtramalo rucyo dipno viṣahārakaḥ. (Bha. Pra.)

Alexiteric, antipyretic, appetiser, cooling, digestive, diuretic, laxative, stomachic.

***Medicinal uses :*** Biliousness, blood diseases, burning sensation, bronchitis, hallucination, leprosy, leucorrhoea, piles.

***Ethno- Medicinal uses :*** Leaves - Jaundice (Tripathi *et al.,* 1996); boils (Deokule, 2006).

Seeds - Cough & cold (Joshi *et al.,* 1980).

Roots - Colic, eczema, gonorrhoea (Shah *et al.,* 1983); snake bite (Goel *et al.*, 1984); diarrhoea (Kataria & Gauchan, 2006); dysentery (Kataria & Gauchan, 2006; Das *et al.,* 2006); skin diseases (Das *et al.,* 2006).

Whole plant - Blood purifier, piles (Sharma *et al.,* 1979); anthelmintic (Jain & Tarafder, 1970).

## *AMORPHOPHALLUS CAMPANULATUS* (Roxb.) Blume ex Decne (ARACACEAE)

***Vern.:*** Jangali suran.

***Regional and Other Names :*** *Bengali* – Ol; *English* - Elephant foot yam; *Gujarati* – Zimikand, jangali suran; *Hindi* – Jangali suran, Jimikand; *Kannada* – Suvarnagedda; *Malayalam* – Chena; *Marathi* – Suran; *Sanskrit* – Surana; *Tamil* - Senai kizhangu; *Telugu* - Kanda dumpa.

***Distribution :*** Cultivated largely throughout the plains of India and Ceylon.

***Taxonomic description :*** A perennial herb. Tubers depressed-globose, bulbiferous, dark brown. Leaves very large, arising after the flowers, long with pale blotches, stout; leaflets obovate; flowers in long spadix.

***Flowering & Fruiting :*** April – May.

***Medicinal parts :*** Corm.

***Taste :*** Acrid – Pungent.

***Chemical constituents :*** Corm contains Proteins, fats, Carbohydrate, Minerals, Calcium Oxalate, Phosphorous, Iron, vitamin A & B.

***Medicinal properties :***

सूरणो दीपनो रूक्षः कषायः कण्डुकृत् कटुः ।
विष्टंभि विशदो रूच्यः कफार्शः कृन्तनो लघुः ।। (भा.प्र.)

Sūṛṇo dīpno ṛūksaḥ kasāyaḥ kaṇdukṛt katuḥ
Viṣtambhi viśado rucyaḥ kaphāṛśaḥ kṛṇtano laguḥ. (Bha. Pra.)

Corm - Acrid, anthelmintic, appetiser, carminative, digestive, expectorant, pungent, stimulant, stomachic.

***Medicinal uses :*** Corm - Asthma, colic, cough, dysentery flatulence, gastric trouble, piles, rheumatism, vomiting, worms.

***Ethno- Medicinal uses :*** Corm - Piles (Shah G.L., 1984), Scabies (Singh *et al.*, 1997); Rheumatic pain (Jain & Sahu, 1993).

## *ANACARDIUM OCCIDENTALE* L. (ANACARDIACEAE)

***Vern.:*** Kaju.

***Regional and Other Names :*** *Bengali* - Kaju, Hajli – badam; *English* - The cashewnut; *Gujarati* – Kaju; *Hindi*–Kaju; *Kannada* - Gera – bija, Gerumara; *Malayalam* - Kappa– Mavu; *Marathi* –Kaju; *Sanskrit* – Kajutaka; *Tamil* – Mundivi; *Telugu* - Jidi anti, Jaedimamidi.

***Distribution :*** A native of tropical America naturalized and cultivated in India.

***Taxonomic description :*** A small tree; leaves coriaceous, obovate-oblong or elliptic, rounded at the apex, glabrous; flowers white, tuning red with age, in terminal panicles; nuts reniform.

***Flowering & Fruiting :*** January– March.

***Medicinal parts :*** Root, bark, fruits.

***Taste :*** Acrid.

***Chemical constituents :*** Nut contain Polysaccharides (Galactose), Galact, Uronic acid.

Leaves contain P–Hydroxybenjoc acid, Protocatechuic acid, Gentistic acid, Gallic acid, Glucosides, Rhamnosides, Arabinosides, Xylosides, Kaempferol, Quercitol.

***Medicinal properties :***

काजूतकस्तुतुवरो मधुरोष्णोलघुः स्मृतः ।
धातुवृद्धिकरोवातकफगुल्मोदरज्वरान् ।।
कृमिव्रणाग्निमांधानिकुष्ठाचश्वेतकुष्ठकम् ।
संग्रहण्यर्शानाहान्नाशयेदितिकीर्तितः ।। (नि. र.)

Kājutakaṣtutuvaro madhuroṣṇolaghuḥ smṛtaḥ
Dhātuvṛiddḥikarovāt kaphaguḷmoḍarajvarāṇ
Kṛmivṛaṇāgṇimāndhānikuṣthācaśvetakuṣthakaṃ
Sangrahaṇyarśānāhāṇnāsayeditikiṛtitaḥ. (Ni. Ra.)

Root - Puragtive.

Fruit - Antidiarrhoeal, emollient, nutritive, rubefacient.

Stem bark - Alterative, astringent.

***Medicinal uses :*** Fruit - Dropsy, dyspepesia, fever, kidney trouble, piles.

Bark - Leprosy, ulcer.

***Ethno- Medicinal uses :*** Bark - Diarrhoea (Gopakumar *et al.*, 1991); ringworm & ulcer (Gupta *et al.,* 1997).

Seed- Leprosy, psoriasis (Rajendran & Rengamani, 2006).

## *ANACYCLUS PYRETHRUM* DC. (ASTERACEAE)

***Syn.:*** *A. officinarum* Hayne.; *Anthemis pyrethrum* DC.

***Vern.:*** Akarkara.

***Regional and Other Names :*** *Bengali* – Akarkara; *English* – Spanish Pellitory; *Gujarati* – Akorkaro; *Hindi* – Akarkara; *Marathi* – Akkalkara; *Sanskrit* – Akarkarabh.

***Distribution :*** A native of Arabia naturalized and cultivated in India.

***Taxonomic description :*** A perinnial procumbent herb; leaves bipinnately compound; flowers yellow in heads.

***Medicinal parts :*** Root, Leaves, Flowers and Fruits.

***Taste :*** Bitter.

***Chemical constituents :*** Alkaloids, Protein, Sugars, Volatile oils, Polyacetylenic compounds, Anacyclin, Hydrocarbolin, Inulin, Pellitorine, Sesamin, Tyramineamids.

***Medicinal properties :***

अकल्लकोष्णो वीर्येण बलकृत् कटुको मतः ।
प्रतिश्यायं च शोथं च वातं चैव विनाशयेत् ।। (शालि. नि.)

Akaḷlakoṣṇo vīryeṇa balkṛt katuko mataḥ
Pṛatiśyāyaṃ ca śothaṃ ca vātaṃ caiva vināśayeṭ. (Sha. Ni.)

Anti-bacterial, Sedative, Stimulant.

***Medicinal uses :*** Epilepsy, paralysis, rheumatism, sexual disorders, toothache.

***ANDROGRAPHIS PANICULATA*** (Burm. f.) Wall. ex Nees (ACANTHACEAE)

***Syn.:*** *A. subspathulata* Clarke; *Justicia Paniculata* Burm.

***Vern.:*** Kalmegh.

***Regional and Other Names :*** *Bengali* – Kalmegh; *English* – Creat; *Gujarati* – Kariyatu; *Hindi* - Kalmegh; *Kannada* – Nelaberu; *Malayalam* – Nelavepu; *Marathi* –Pelekiraiet; *Sanskrit* – Bhunimba ; *Tamil* – Nelavembu; *Telugu* –Neelavemu.

***Distribution :*** Throughout India.

***Taxonomic description :*** Erect, glabrous, annual. Branches 4-angled. Leaves slightly undulate; base tapering. Flowers small, solitary, whitish, spotted with rose-purple. Corolla 2-lipped. Capsules linear-oblong.

***Flowering & Fruiting :*** October-December.

***Medicinal parts :*** Whole plant, leaves, roots.

***Taste :*** Bitter.

***Chemical constituents :*** Plant contains Kalmeghin, Andrographolid, Tannic acid, Sodium and Potassium.

***Medicinal properties :***

किरातः सारको रूक्षः शीतलस्तिक्तो लघुः ।
सन्निपातज्वर श्वास कफपित्तास्त्रदाहनुत् ।। (भा. प्र.)

Kirātaḥ sārako ruk̇ṣaḥ śītalaṣtiḳto laghuḥ
Saṇnipātajvara śvāsakaphapịttāṣradāhanuṭ. (Bha. Pra.)

Antiperiodic, digestive, Hepatic antileprotic, nutritive, stimulant, stomachic.

***Medicinal uses :*** Anorexia, blood purifier, hepatomegally, malaria, oedema, skin diseases.

***Ethno- Medicinal uses :*** Whole plant - Dysentery, diarrhoea, fever (Nath *et al.*, 2006); anthelmintic (Srivastava & Varma, 1981; Goel & Mudgal, 1988); fever (Jain & De, 1966); filaria, dyspepsia, dysentery (Sudhakar & Rolla, 1985); gastric complaint (Goel *et al.*, 1984; Gupta, 1981); indigestion (Tarafder, 1986); liver complaint (Hemadri & Rao, 1984); skin diseases (Jain, 1963, 1965; Rajendran & Rengamani, 2006); snake bite (Jain, 1963, 1973).

Leaves - Fever (Purkayastha & Subhan, 2006; Das *et al.,* 2006); snake bite (Ganesan *et al.*, 2006); malaria (Deka *et al.,* 2006; Gupta, 2006); jaundice (Kadavul & Jisha, 2006; Dutta Choudhury & Shil, 2006); dysentery (Das *et al.,* 2006); stomach trouble (Das *et al.,* 2006); anthelmintic (Das *et al.,* 2006).

Root - Dysentery, fever, liver complaint, malaria, dyspepsia (Bhalla *et al.,* 1982).

Stem – Tuberculosis (Mokat & Deokule, 2006); dysentery (Das *et al.,* 2006); anthelmintic (Das *et al.,* 2006).

## *ANNONA SQUAMOSA* L. (ANNONACEAE)

***Vern.:*** Seetaphaal.

***Regional and Other Names :*** *Bengali* – Seetaphal; *English* - Custard apple; *Gujarati* – Anan; *Hindi* –Seetaphal; *Kannada* – Seethaphala; *Malayalam* - Seetha pazham; *Marathi* – Seetaaphala; *Sanskrit* – Seetaphala; *Tamil* - Sethapazham, atta; *Telugu* – Seethapalamu.

***Distribution :*** A native of West- Indies, Now cultivated throughout India.

***Taxonomic description :*** Evergreen shrubs or small trees with oblong-lanceolate leaves and yellowish green flowers. Fruits globose or ovoid with projecting areoles.

***Flowering & Fruiting :*** July – November.

***Medicinal parts :*** Roots, Leaves, fruits & Seeds.

***Taste :*** Sweet.

***Chemical constituents :*** Leaves contain Anonaine, Roemerine, Corydine, Narcorydine, Norlaureline, Isocorydine, Noricocorydine, Glucine, Friedeline, Alkaloids, Essential oils, Volatile oil, Carvone, Michelabine, Liriodedine, Diazepine, Squamolone, Beta–Caryophylline, β-sitosterol, Camphor, Barmeol.

***Medicinal properties :***

सीताफलं तु मधुरं शीतं हृदयं बलप्रदम् ।
वातलं कफकृत् स्वादु पुष्टिकृत्पित्तनाशनम् ।। (नि. र.)

Sītāphalalṃ tu madhuraṃ sītaṃ hṛdyaṃ balapṛadaṃ
Vātalaṃ kaphakṛt ṣvādu puṣtikṛt pittanāsanaṃ. (Ni. Ra.)

Root - Purgative.

Bark - Astringent.

Leaves - Anthelmintic, antispasmodic, stimulant, sudorific.

Fruits - Antiscorbutic, astringent, Cooling, febrifuge, pectoral.

Seeds - Abortifacient, emetic.

***Medicinal uses :*** Bark - Diarrhoea.

Leaves - Boils, swelling, ulcers.

Fruits- Diarrhoea, dysentery, dyspepsia, vertigo, vomiting.

***Ethno- Medicinal uses :*** Leaves - Dandruff (Basak, 1997).

Fruits - Skin diseases (Upadhay *et al.,* 1998).

Seed- Dandruff (Arinathan *et al.,* 2006).

## *ANOGEISSUS LATIFOLIA* (Roxb. ex DC.) Wall. ex Guill. & Perr. (COMBRETACEAE)

***Vern.:*** Dhawra.

***Regional and Other Names :*** *Bengali* – Dhaoyo; *English* - Button tree; *Gujarati* – Dabria; *Hindi* –Dhawra; *Kannada* – Diadinduga; *Malayalam* – Vellanava; *Marathi* –Dhaura; *Sanskrit* – Dhava; *Tamil* – Vekkali; *Telugu* – Vellama.

***Distribution :*** Throughout the greater part of India.

***Taxonomic description :*** Tall, deciduous trees, with smooth, white-grey bark. Leaves silky tomentose when young. Flowers 2-5 in a head, yellow to pinkish- yellow. Fruits orbicular, glabrous, with entire wings.

***Flowering & Fruiting :*** October – June.

***Medicinal parts :*** Root, bark, fruits.

***Taste :*** Bitter.

***Chemical constituents :*** Seed contain Protein.

Leaves contain Phenol, Carboxylic acid, Saponin, Amino acids, Sugars.

Stem bark contain Shikimic acid, quinic acid, fructose, alanine, phenylalanine, tannins. Saponins.

Sap wood contain Quinic acid, glucose, phenylalanine, ellagic acid.

Heart wood contain Quercetin, myricetin, glucose, ellagic acid, gallic acid, phenylalanine.

***Medicinal properties :***

धवः शीतः प्रमेहार्शः पाण्डुपित्तकफापहः । (भा.प्र.)

Dhavaḥ śītaḥ pramehārśḥ paṇdupittakaphāpahaḥ. (Bha.Pra.)

Root - Acrid, pungent, stomachic.

Bark - Acrid, appetiser, cooling, pungent, sweet.

***Medicinal uses :*** Bark - Anaemia, erysepelas, piles, skin diseases, urinary discharges.

***Ethno- Medicinal uses :*** Root - Headache (Gopal & Shah, 1985); Stomach diseases (Shah *et al.*, 1981).

Gum - Dysentery (Roy & Chaturvedi, 1986); Debility after pregnancy (Jadhav, 2006).

Bark - Headache (Shah *et al.*, 1981); Cough & Stomachic (Ramachandran & Nair, 1981); Snake & Scorpion bite (Mishra & Billore, 1983); Colic (Apparanantham & Chelladurai, 1986); Diarrhoea & dysentery (Maheshwari *et al.*, 1986); skin diseases (Kumar *et al.*, 2006).

## *ANTHOCEPHALUS CADAMBA* (Roxb.) Miq. (RUBIACEAE)

***Syn.:*** *A. chinensis* Lam; *A. indicus* A. Rich.

***Vern.:*** Kadam.

***Regional and Other Names :*** *Bengali* – Kadam; *English* – Kadam; *Gujarati* – Kadamb; *Hindi* – Kadamb; *Kannada* – Kaduavalatige; *Malayalam* – Katampu; *Marathi* – Kadam; *Sanskrit* – Kadambah; *Tamil* – Vellaikkatampu; *Telugu* – Rudrakskamba.

***Distribution :*** From the Himalaya to Ceylon and Malacca.

***Taxonomic description :*** A large tree with spreading branches; leaves elliptic-oblong, glabrous above; flowers sweet scented, fruits fleshy.

***Flowering & Fruiting :*** June – September.

***Medicinal parts :*** Bark, leaves, fruits.

***Taste :*** Bitter, astringent.

***Chemical constituents :*** Leaves contain Alkaloid (Kadambine & 3 – œ-di – hydrocadambine); Glycosides alkaloids (Iso – dihydrocadambine & β-Sitosterol).

Root & stem bark contain Quinovic acid, Pentacyclic triterpenic acid, cadambagenic, quinioric acid, saponin.

Flower contain Essential oils.

***Medicinal properties :***

कदम्बो मधुरः शीतः कषायो लवणो गुरुः ।
सरोऽवष्टम्भकृद्रूक्षः कफास्तन्यानिलप्रदः ।। (भा.प्र.)

Kadaṃbo madhuraḥ śītaḥ kasāyo lavano guruḥ
Saroavaṣtaṃbhakṛdṛūksaḥ kaphāṣtaṇyānilapṛadaḥ. (Bha. Pra.)

Acrid, anodyne, astringent, anti – inflammatory, bitter, carminative, constipation cooling, digestive, diuretic, expectorant, febrifuge.

***Medicinal uses :*** Bark - Cough, debility, diarrhoea, dyspepsia, fever, inflammation, ulcers, vomiting, wound.

Leaves - Ulcers, wounds.

Fruits - Fever.

***Ethno- Medicinal uses :*** Bark - Antifertility (Tarafder, 1983); fever (Sebastian & Bhandari, 1984); snakebite (Karnich *et al.,* 1981); syphilis (Kumar & Jain, 1998).

## *ARGEMONE MEXICANA* L. (PAPAVERACEAE)

***Vern.:*** Satyanashi.

***Regional and Other Names :*** *Bengali* – Shialkanta; *English* - Mexican poppy; *Gujarati* – Darudi; *Hindi* –Satyanashi; *Kannada* – Datturigidda; *Malayalam* – Bhrahmadande; *Marathi* – Pilvadhotra; *Sanskrit* –Svarnaksiri; *Tamil* – Bhrahmadandu; *Telugu* – Brahmadandi.

***Distribution :*** Introduced and naturalized throughout India. Indigenous in tropical America.

***Taxonomic description :*** Prickly herbs, with yellow sap and pinnatifid and spiny leaves. Flowers yellow, solitary terminal. Capsule prickly.

***Flowering & Fruiting :*** February-June.

***Medicinal parts :*** Root, yellow latex.

***Taste :*** Bitter.

***Chemical constituents :*** Root contains Berbetine, Protopine nitrate, Betasterol, fatty acids, ceryl alcohol, Allocryptopine.

Flowers contain Tannin, Resin, Argemone oil.

Leaves contain Glucosides, Kampferin, Anthraquinone, essential oil & Chrysophanic acid.

Seeds contain tasteless oil of dark brown colour, Carbohydrates, Albumin, Phosphate, sulphate.

***Medicinal properties :***

स्वर्णक्षीरी हिमा तिक्ता क्रिमिपित्तकफापहा
मूत्रकृच्छ्राश्मरीशोफदाहज्वरहरा परा ।। (रा. नि.)

Ṣvaṛnakṣīrī himā tiktā kṛimipittakaphāpahā
Muṭrakṛacchāśmarīśophadāhajvaraharā parā. (Ra. Ni.)

Root - Alterative, anodyne, Diuretic.

Seed - Demulcent, emetic, expectorant, laxative.

***Medicinal uses :*** Root - Skin diseases.

Seed - Diarrhoea, dysentery.

***Ethno- Medicinal uses :*** Root - Eye complaint (Bhalla *et al.*, 1982); Skin diseases (Singh & Maheshwari, 1983); Piles (Kamble & Pradhan, 1980); Dog bite (Pal, 1980); Scabies (Sudhakar & Chetty, 1998), Leucoderma (Singh *et al.,* 1997); stomachache, vomiting (Subhangi & Patil, 2006); fertility (Upadhyay & Jaswant Singh, 2005); cuts & wounds (Das *et al.,* 2006).

Seed - Cough, asthma, whooping cough (Bhalla *et al.,* 1982); Skin diseases (Shekhawat & Anand, 1984; Kumar & Chauhan, 2005); Boils & ulcer (Sudhakar & Rolla, 1985); Headache (Singh & Pandey, 1980); Gum trouble (Bhatt & Sabnis, 1987).

Latex - Eye complaints (Maheshwari & Singh, 1984); Skin diseases (Shah, 1984; Pandit *et al.*, 1996; Gupta *et al.*, 1997; Deokule, 2006); Toothache (Singh *et al.*, 1987); Scabies (Jain *et al.*, 1973; Pandit *et al.* 1996; Gupta, *et al.* 1997); Jaundice (Shah, G.L. 1984; Pandit *et al.*, 1996; Gupta *et al.*, 1997).

Leaves - Skin diseases, dropsy (Bhalla *et al.,* 1982; Dutta Choudhury & Shil, 2006); Boil, ulcer (Goel *et al.,* 1984); Eye complaint, Scabies (Jain *et al.*, 1973); Leprosy (Shah *et al.*, 1981); Syphilis (Bhatt & Sabnis, 1987); Ringworm (Jain *et al.,* 1973; Jadhav, 2006); jaundice (Bhalla *et al.,* 1982; Dutta Choudhury & Shil, 2006; Das *et al.,* 2006).

## *ARGYREIA SPECIOSA* Sweet. (CONVOLVULACEAE)

***Syn.:*** *A. nervosa* (Burm. f.) Bojer.

***Vern.:*** Samudra sosh.

***Regional and Other Names :*** *Bengali* – Bichtarak; *English* - Elephant creeper; *Gujarati* - Samudra sosh; *Hindi* – Samunder kapat; *Kannada* – Chandrapada; *Malayalam* – Samudrapala; *Marathi* – Samudrasoka, Vriddhddaru; *Sanskrit* - Samudra sosha; *Tamil* – Samuthrappalai; *Telugu* – Chandrapada.

***Distribution :*** Throughout India.

***Taxonomic description :*** White tomentose twiners, with ovate-cordate leaves, tomentose beneath. Flowers rose-purple and white.

***Flowering & Fruiting :*** August – December.

***Medicinal parts :*** Whole plant, root, leaves, seeds.

***Taste :*** Bitter.

***Chemical constituents :*** Plant contains Arginine, Lycine, Valine, Behenic acid, Linolic acid, linolenic acid, Oleic acid, Argoclavine, Clanoclavine–I & II, Elymoclavine, Ergine, Isoergine, Ergometrine, fastuclavine, Isosetoclavin, setoclavine, Lysergine, Lysergol, Lysergic acid, Caffeic acid, Anthocyanin, Cumarine, Flavonoids, Quercetine, Kaempferol, β-Setosterol.

***Medicinal properties :***

वृद्धद्धारुः कषायोष्णः कटुस्तिक्तो रसायनः ।
वृष्यो वातामवातार्शः शोथमेहकफप्रणुत् ।
शुक्रायुर्बलमेधाऽग्निस्वरकान्तिकरः सरः ।। (भा.प्र.)

Vrīḍdhḍdāruḥ kasāyoṣnaḥ katuṣtikto rasāyanaḥ
V̩rīṣhyo vātāmvātā̩rśhaḥ śothmehkaphap̩ra̩nu̩t
Śuk̩rāyu̩rbalmeghāagniṣvarkā̩ntikaraḥ saraḥ. (Bha. Pra.)

Leaf - Emollient.

Root - Alterative, bitter, diuretic.

***Medicinal uses :*** Leaf - Eczema, itching, ringworm, skin diseases.

Root - Gonorrhoea, rheumatism, syphilis.

***Ethno- Medicinal uses :*** Leaf - Boils (Pal, 1980); headache (Saxena & Dutta, 1975); swelling (Deokule, 2006); Gout (Jadeja *et al.,* 2006).

Root - Diabetes, gonorrhoea & ringworm (Tribedi *et al.,* 1982).

## *ARISTOLOCHIA INDICA* L. (ARISTOLOCHIACEAE)

***Syn.:*** *A. lanceolata* Wight.

***Vern.:*** Isarmul.

***Regional and Other Names :*** *Bengali* – Isarmul; *English* - Indian-birthwort; *Gujarati* – Arkmula; *Hindi* – Isarmul; *Kannada* – Eashwari-beru; *Malayalam* – Ishwaramulli; *Marathi* – Sapasan; *Sanskrit* – Sampsun, Nakuli; *Tamil* – Isuramuli; *Telugu* –Dulagovela.

***Distribution :*** W. Peninsula, Ceylon.

***Taxonomic description :*** Twining, woody herbs, with elliptic-ovate or obovate- panduriform leaves. Flowers greenish-white, in axillary racemes. Capsules oblong-globose; seeds winged.

***Flowering & Fruiting :*** September – November.

***Medicinal parts :*** Whole plant.

***Taste :*** Bitter.

***Chemical constituents :*** Plant contains Three bases of Arisotolochine, Nitrogenous acids (Aristinic acid, Aristidinic acid & Aristolic acid); Volatile oil.

***Medicinal properties :***

नाकुली तुवरा तिक्ता कटुकोष्णा विनाशयेत्।
भोगिलूतावृश्चिकाखुविषज्वरकृमिव्रणान्।। (भा. प्र.)

Nākulī tuvarā tiktā katukoṣnā vināśyeṭ
Bhogilūtāvṛiścakākhuvisjvarkṛimivṛnāṇ. (Bha. Pra.)

Analgesic, anthelmintic, anti-inflammatory, antipyretic, cardiotonic, purgative, stimulant, stomachic, tonic.

***Medicinal uses :*** Amenorrhoea, erysepelas, fever, flatulence, rheumatism, scorpion sting, snakebite, spider bite, strangury, ulcer, worms.

***Ethno- Medicinal uses :*** Leaf - Abortifacient (Tarafder, 1983); snake bite (Nagendra Prasad & Abraham, 1984).

Stem - Skin diseases (Chaudhuri & Pal, 1976).

Seed - Arthritis (Tripathi *et al.,* 1996).

Root - Stomach troubles (Tarafder, 1983); abortifacient (Tarafder, 1983); snake bite (Sudhakar & Rolla, 1985); skin diseases (Ramachandra & Nair, 1981); scabies (Mishra & Billore, 1983); wounds & diarrhoea (Saxena *et al.,* 1981); leucoderma (Basak, 1997).

## *ASPARAGUS RACEMOSUS* Willd. (LILIACEAE)

***Syn.:*** *A. javanica* Kunth.; *A. sarmentosus* Graham.

***Vern.:*** Shataveri.

***Regional and Other Names :*** *Bengali* – Satmuli; *English* – Asparagus; *Gujarati* –Satavari; *Hindi* – Satawar; *Kannada* - Majjige-gadde aheruballi; *Malayalam* – Satavalli; *Marathi* -Satmuli; *Sanskrit* – Shatapadi, Satavari; *Tamil* – Ammaikodi; *Telugu* –Challagadda.

***Distribution :*** Throughout India.

***Taxonomic description :*** Scandent, spinous undershrubs, with faciculated tuberous roots. Leaves absent, branches modified into linear, leaf-like cladodes. Flowers in racemes. Fruits globose.

***Flowering & Fruiting :*** September – December.

***Medicinal parts :*** Fleshy roots.

***Taste :*** Sweet.

***Chemical constituents :*** Fleshy roots contain Sugars, mucilage & saponins.

***Medicinal properties :***

शतावरी गुरुः शीता तिक्ता स्वाद्वी रसायनी ।
मेधाग्नि पुष्टिदा स्निग्धा नेत्रया गुल्मातिसारजित् ।। (भा. प्र.)

Śatāvarī guruḥ śītā tiktā ṣvāḍvī rasāyanī
Medhāgni puṣtidā ṣnigdhā neṭryā guḷmātisārajiṭ. (Bha. Pra.)

Roots - Alterative, aphrodisiac, appetiser, astringent, Bitter, cooling, indigestion, stomachic, sweet, tonic.

***Medicinal uses :*** Root - Biliousness, diseases of the blood & the eye, dysentery, epilepsy, inflammation, leprosy, night blindness, throat complaint, tuberculosis, tumour.

***Ethno-medicinal uses*** : Root - Rheumatism (Lakshamanan & Sankranarayanan, 1988); cough (Sebastian & Bhandari, 1984); fertility problem, to increase milk of mother (Mokat & Deokule, 2006); dysentery & wound (Chaudhuri *et al.,* 1975); diarrhoea (Sharma *et al.,* 1974); gastric complaint (Joshi *et al.,* 1980); gonorrohoea (Tarafder, 1984); leucorrhoea (Hemadri & Rao, 1983); Impotency (Ramchandran & Nair, 1981); stomach trouble (Hajra & Baishya, 1981); dyspepsia & anaemia (Gopakumar, 1991); general debility (Singh & Aswal, 1992; Kumar & Chauhan, 2005); typhoid (Jadhav, 2006); improve sperm count (Udayan *et al.*, 2006 b); retention of urine (Kumar & Chauhan, 2005); tonic (Deokule, 2006); fever (Baruah *et al.*, 2006); dysentery (Rajendran & Rengamani, 2006).

Flower- Gout (Jadeja *et al.*, 2006).

## *AZADIRACHTA INDICA* A. Juss. (MELIACEAE)

***Syn.:*** *Melia azadirachta* L.

***Vern.:*** Neem.

***Regional and Other Names :*** *Bengali* – Nim; *English* - Margosa tree; *Gujarati* – Limba; *Hindi* –Neem; *Kannada* – Bevu; *Malayalam* – Veppa; *Marathi* – Limba; *Sanskrit* – Nimbaka; *Tamil* – Vembu; *Telugu* – Veepachettu.

***Distribution :*** Throughout India.

***Taxonomic description :*** Evergreen trees, with serrate or crenate, 7-17 leaflets and white flowers in panicles. Drupes ovoid, yellow when ripe.

***Flowering & Fruiting :*** March – May.

***Medicinal parts :*** Whole plant, root bark, stem bark, leaves, flowers, fruits.

***Taste :*** Bitter.

***Chemical constituents :*** Plant contains Margocin, Nimbadin, Sulphur, Nimban, Nimbosterol, Tannin, essential oil, Resin, Glycosides, Fatty acids, gum, free amino acids.

Leaves contain Quercetin, β- Sitosterol (Nimbasterol).

***Medicinal properties :***

निम्बः शीतोलघुर्ग्राही कटुपाकोग्निवातनुत् ।
अहृद्यः श्रमतृड्कास ज्वरारुचिक्रिमिप्रणुत् ।। (भा.प्र.)

Niṃbaḥ śīto laghuṛgrāhī katupākognivātanuṭ
Aḥṛdyah śramatṛtikāsa jvrārucikṛimipṛaṇuṭ. (Bha. Pra.)

Stem bark - Anthelmintic, astringent, Bitter, demulcent, purgative, tonic.

Leaves - Anthelmintic, anti – inflammatory, antileprotic, stimulant.

***Medicinal uses :*** Stem bark - Burning sensation, fever, itching, skin diseases, Ulcer, urinary disorders.

Leaves - Diseases of the eye, fever, inflammation, skin diseases, ulcer, urinary disorders.

***Ethno- Medicinal uses :*** Leaf - Boils & wounds (Shah, 1982); cold & diabetes (Shah, 1984); diarrhoea & dysentery (Kharkongor & Joseph, 1981); blood purifier (Rao, 1981); gout (Sebastian, 1984); heart complaint (Rao, 1981); scabies (Nagendra Prasad & Abraham, 1984); skin diseases (Rai, 1985; Pereira, 2006; Deokule, 2006; Kumar *et al.,* 2006; Rajendran & Rengamani, 2006; Basak, 2006; Das *et al.,* 2006); piles (Singh, 1988); malaria (Shekhawat & Anand, 1984; Deka *et al.,* 2006); snake bite (Lal & Yadav, 1983; Jain & Sahu, 1993); Ascariasis (Jadhav, 2006); mumps (Ganeshan *et al.*, 2006); dyspepsia (Jadeja *et al.,* 2006); itches (Sarma *et al.,* 2006); Hepatitis (Dutta Choudhury & Shil, 2006); diabetes (Ghosh, 2006; Das *et al.,* 2006); small pox (Rajendran & Rengamani, 2006); eczema, ulcer (Basak, 2006); ulcer (Das *et al.,* 2006).

Flower - Dyspepsia (Bhalla *et al.,* 1982); wound (Maheshwari & Singh, 1984); malaria (Deka *et al.,* 2006).

Fruits - Leprosy & skin diseases (Shah, 1982); piles (Bhalla *et al.,* 1982).

Seed - Antiseptic (Sudhakar & Rolla, 1985); skin diseases (Singh & Pandey, 1980; Singh & Prakash, 1994); malaria (Deka *et al.,* 2006).

Seed oil - Piles (Ved prakash & Jain, 1984); skin diseases (Shah, 1984); ulcer (Jain & De, 1966).

Bark - Jaundice & liver complaint (Bhalla *et al.,* 1982); wounds (Shekhawat & Anand, 1984; Deokule, 2006); skin diseases (Shekhawat & Anand, 1984; Deokule, 2006; Kumar *et al.,* 2006; Basak, 2006); fever (Kumar *et al.,* 2006; Arinathan *et al.,* 2006); malaria (Deka *et al.,* 2006); cold (Arinathan *et al.,* 2006); swelling (Sharma *et al.,* 2006); ulcer, eczema (Basak, 2006).

Gum - Snake bite & scorpion sting (Jain & Sahu, 1993); diarrhoea (Mohanty *et al.,* 1996).

## ***BACOPA MONNIERI*** (L.) Penn. (SCROPHULARIACEAE)

***Syn.:*** *Herpestis monnieri* (L.) H.B. & K.; *Moniera cuneifolia* Michx.; *Gratiola monniera* L.

***Vern.:*** Brahmi.

***Regional and Other Names :*** *Bengali* - Brahmi sak; *English* - Thyme leaved-gratiola; *Hindi* - Brahmi; *Kannada* – Nirubrahmi; *Malayalam* – Neerbrahmi; *Marathi* – Neerbrahmi; *Sanskrit* - Brahmi; *Tamil* - Neerbrahmi; *Telugu* – Sambranichettu.

***Distribution :*** Frequent in marshy & swampy places.

***Taxonomic description :*** A prostrate gregarious herb with spreading or ascending branches, rooting at the nodes. Leaves sessile, decussate, obovate-oblong or spathulate, succulent. Flowers axillary, solitary on slender pedicels. Corolla bluish –purplish; anthers bluish-purple. Capsules ovoid, glabrous.

***Flowering & Fruiting :*** June – October.

***Medicinal parts :*** Whole plant.

***Taste :*** Bitter.

***Chemical constituents :*** Plant contains Alkaloids – Brahmine & Herpestine, Bacoside A & B (Saponins), Betulic acid, Hersaponin, Nicotine, Becogenin – A, Potassium, Sodium, Calcium, Iron, Oleic acid, Palmitic acid.

***Medicinal properties :***

ब्राह्मी हिमासरातिक्ता लघुर्मेध्या च शीतला ।
कषाया मधुरा स्वादुपाकाऽऽयुष्या रसायनी ।। (भा. प्र.)

Bṛāḥmī himāsarātiktā laghuṛmeḍhyā ca śītalā
Kasāyā madhurā ṣvādupākāaayusyā rasāyanī. (Bha. Pra.)

Anti – inflammatory, blood purifier, diuretic, laxative, tonic.

***Medicinal uses :*** Blood diseases, cough, debility after heart attack, dyspepsia, epilepsy, fever, hoarsness of voice, insomnia, less memory, tension.

***Ethno-Medicinal uses :*** Leaf - Abdominal pain (Joshi, 1982); cough (Malhotra & Moorthy, 1973); liver complaint (Bhatt, & Sabnis, 1987); rheumatism (Shah *et al.,* 1983).

Whole plant - Blood purifier, boils (Uniyal *et al.*, 1973); fever, headache (Lakshmanan & Sankaranarayanan, 1988); eczema, ringworm (Saxena & Vyas, 1983); stomach complaint (Shah *et al.*, 1981); nervous disorders, mental diseases, hoarsness (Das, 1995; Tripathi *et al.*, 1996); snake bite (Jain & Sahu, 1993); asthma (Basak, 2006); epilepsy (Uniyal *et al.*, 1973; Basak, 2006).

## *BALANITES AEGYPTIACA* (L.) Del. (SIMAROUBIACEAE)

***Syn.:*** *B. rouxburghii* Planch.; *B. ingudi*; *Ximenia aegyptiaca* L.

***Vern.:*** Hingot.

***Regional and Other Names :*** *Bengali* – Hingon; *English* - Desert date; *Gujarati* - Hinger; *Hindi* - Hingot; *Malayalam* – Nanjunta; *Marathi* – Hingan; *Sanskrit* -Inguda; *Tamil* - Nanjundan; *Telugu* – Gara.

***Distribution :*** Drier parts of India.

***Taxonomic description :*** Spinous trees or shrubs, with 2-foliolate leaves. Flowers greenish-yellow, in axillary fascicled cymes. Fruits oblong, rind hard.

***Flowering & Fruiting :*** March – June.

***Medicinal parts :*** Fruits, seeds.

***Taste :*** Bitter (Root), Sweet (Fruit).

***Chemical constituents :*** Fruits & seeds contain Oil of yellow colour and tasteless.

Pulp & fruit contains Saponin.

***Medicinal properties :***

इंगुदः कुष्ठभुतादिग्रहव्रणविषक्रिमीन् ।
हन्त्युष्णः श्वित्रशूलघ्नस्तिक्तकः कटुपाकवान् ।। (भा. प्र.)

Iṇgudaḥ kuṣthabhutadigrahavṛanavisakṛimīṇ
Haṇtyuṣṇaḥ svitrasulaghnaṣtiktaḥ katupakvaṇ. (Bha. Pra.)

Flower - Bitter, heating, sweet smelling.

Fruits - Antidysenteric, alexipharmic, alterative, analgesic anthelmintic, bitter sharp digestive, expectorant, laxative, taste.

Leaf - Anthelmintic, bitter, laxative.

***Medicinal uses :*** Fruits - Cough, leprosy, skin diseases, ulcer, wound, worm infestation.

Leaf - Cough, throat infection.

***Ethno- Medicinal uses :*** Seed oil - Burn & cough (Shah, 1981).

Bark - Cold (Bhalla *et al.,* 1982); cough (Shekhawat & Anand, 1984); swelling (Maheshwari *et al.,* 1986); snake bite (Deokule, 2006).

Fruits - Cold & cough (Bhalla *et al.,* 1982); pneumonia (Saxena & Vyas, 1983); snake bite (Singh, 1945); skin diseases (Shekhawat, 1984; Pandit *et al.,* 1996; Sharma *et al.,* 2006); dysentery (Jadeja *et al.,* 2006); boils, leucoderma, whooping cough (Sharma *et al.,* 2006).

Leaves- Snake bite (Deokule, 2006).

## *BARLERIA PRIONITIS* L. (ACANTHACEAE)

***Vern.:*** Vajradanti.

***Regional and Other Names :*** *Bengali* - Kanta-jati; *English* - Katsareya; *Gujarati*-Pilikanta-shelio; *Hindi*-Vajradanti; *Kannada* – Kurantaka; *Malayalam* – Shemmuli; *Marathi* – Koranti; *Sanskrit* - Saireyaka; *Tamil* - Chemmuli; *Telugu* - Kanakam-baram.

***Distribution :*** Throughout India.

***Taxonomic description :*** Armed undershrubs, with elliptic or ovate-elliptic leaves and yellow flowers in dense spikes. Calyx-lobes glandular. Capsules 4-seeded.

***Flowering & Fruiting :*** October – January.

***Medicinal parts :*** Root, leaves.

***Taste :*** Bitter.

***Chemical constituents :*** Plant contains Beta- Sitosterol, Scutellarein – 7 Rhamocyl Glucoside, Iridoids, Barlerin and Ac- Barlerin.

***Medicinal properties :***

सैरेयः कुष्ठवातास्त्रकफकण्डुविषापहः ।
तिक्तोष्णो मधुरोऽनमम्लः सुस्निग्धः केशरंजनः ।। (भा. प्र.)

Sareyahaḥ kuṣthavātāṣṭrkaphkaṇduvisāpahaḥ
Tiktoṣṇo madhuroanamṃlahaḥ suṣnigdhaḥ keśhraṇjanaḥ.
(Bha. Pra.)

Acrid, Anti-inflammatory, Antiseptic, Appetiser, bitter, diaphoretic, diuretic, expectorant, heating.

***Medicinal uses :*** Leaf juice- Abdominal disorders, catarrha, fever.

Dried bark - Cough.

Decoction of plant - Cough, dropsy, Paralysis, toothache.

Paste of root - is applied on boils & glandular swelling.

***Ethno- Medicinal uses :*** Root - Fever (Jadhav, 2006); boils (Bhalla *et al.*, 1982; Upadhaỳ *et al.*, 1998); toothache (Basak, 1997); eczema (Deokule, 2006).

Bark - Dropsy (Bhalla, 1982); Gout (Jadeja *et al.*, 2006).

Leaf - Toothache (Sudhakar & Rolla, 1985; Ahmad, 1995); wound (Ahmad, 1995); cough (Baruah & Sarma, 1987; Ahmad, 1995); rheumatism (Saxena & Vyas, 1983); piles (Singh, 1988); ear complaint (Mishra & Dixit, 1973).

Flower- Toothache (Aditya *et al.*, 2006).

## *BAUHINIA PURPUREA* L. (CAESALPINIACEAE)

***Vern.:*** Kaliar.

***Regional and Other Names :*** *Bengali* - Rakta kanchan; *English* - Butterfly tree; *Hindi* - Kaliar; *Kannada* - Sarul; *Malayalam* - Suvannaman-daram; *Marathi* - Atmatti; *Sanskrit* - Kovidar; *Tamil* - Deakanjan; *Telugu* – Devakanjan.

***Distribution :*** Throughout India.

***Taxonomic description :*** Tree; branches stout, glaberescent; leaves roundish, divided half way down into 2 obtuse lobes; flowers fragrant in terminal and axillary few flowered racemes; pods subwoody, flat pointed at both ends, glabrous; seeds 12-16, oblong-ellipsoid, brown.

***Flowering & Fruiting :*** September – November.

***Medicinal parts :*** Stem bark, flowers & leaves.

***Taste :*** Acrid.

***Chemical constituents :*** Flower contains Pelargonidin–3–glucoside, Pelargonidin -3- Triglucoside, Chalcon galactoside, Isoquercetin and Astragalin.

***Medicinal properties :***

कोविदारो हिमो ग्राही कफपित्तजित् ।
गण्डमालागुदभ्रंशव्रणकुष्ठकृमीन् जयेत् ।। (कै. नि.)

Kovidaro himo grahi kasayah kaphpittjit
Gandmalagudbrhnsvrnkusthakrimin. (Kai. Ni.)

Anthelmintic, astringent, carminative, cooling, laxative, rejuvenator, tonic.

***Medicinal uses :*** Flower - Blood dysentery, piles.

Stem bark - Diarrhoea.

Root bark - Cough, goitre, haemorroids, leucoderma.

***Ethno- Medicinal uses :*** Stem - Bone fracture (Kumar *et al.,* 1980).

Flower - Indigestion (Rao & Jamir, 1982); diarrhoea (Rana *et al.*, 1996).

Leaf - Jaundice (Apparanantham & Chelladurai, 1986).

Bark - Small pox (Jain & De, 1966); wound (Karnick *et al.,* 1981; Pereira, 2006); diarrhoea (Rana *et al.*, 1996; Kataria & Gauchan, 2006; Das *et al.,* 2006); dysentery (Chhetri, 2005; Kataria & Gauchan, 2006); leucorrhoea (Das *et al.,* 2006); menstrual trouble (Das *et al.,* 2006):

## *BAUHINIA VARIEGATA* L. (CAESALPINIACEAE)

***Vern.:*** Kachnar.

***Regional and Other Names :*** *Bengali* - Raktakanchan; *English* – Camel's foot; *Hindi* - Kachnar; *Kannada* - Kanchavala; *Malayalam* - Mandarum; *Marathi* -Raktakanchan; *Sanskrit* - Kanchanara; *Tamil* - Taruvatti; *Telugu* - Adavimandara.

***Distribution :*** Throughout India.

***Taxonomic description :*** Medium-sized deciduous trees. Leaves orbicular, 2-lobed. Flowers white to pink. Pods flat, glabrous, brown.

***Flowering & Fruiting :*** August– February.

***Medicinal parts :*** Stem bark & flowers.

***Taste :*** Acrid.

***Chemical constituents :*** Stem bark contains Sugar & Tannins.

Flowers contain Octacosanol, β- stiosterol, flavinoid.

Seeds & bark contain Myrcetol, glycoside, kaempferol, glucoside.

***Medicinal properties :***

काञ्चनारो हिमो ग्राही तुवरः श्लेष्मपित्तनुत् ।
कृमिकुष्ठगुदभ्रंशगण्डमालाव्रणापहः ।। (भा.प्र.)

Kāncaṇārō himo grāhī tuvaraḥ śleṣhmapiṭtanuṭ
Kṛmikuṣthagudbhṛṇśagaṇdmālāvṛanāpaḥ. (Bha. Pra.)

Bark - Alterative, astringent, tonic.

Flowers - Acrid, astringent, cooling, sweet.

***Medicinal uses :*** Bark - Asthma, blood diseases, dysmenorrhoea, leprosy, leucoderma, stomach trouble, tuberculous glands, ulcers, wound.

Flowers - Blood diseases, bronchitis, headache, vaginal discharges.

***Ethno- Medicinal uses :*** Bark - Leprosy, ulcer, skin diseases, syphilis (Bhalla *et al.,* 1982); malaria, snake bite (Negi *et al.,* 1985); dysentery (Megoneitso & Rao, 1983; Kataria & Gauchan, 2006); piles (Rajwar, 1983); tumour (Shah & Joshi, 1971); diarrhoea (Negi *et al.,* 1985; Kataria & Gauchan, 2006); wound, swelling (Kumar *et al.,* 2006).

Flower - Dysentery (Rajwar, 1987; Gopakumar, 1991); worm & diarrhoea (Jain & De, 1966); tumour (Shah & Joshi, 1971); Oedema (Gopakumar *et al.*, 1991); piles (Gopakumar *et al.*, 1991; Kumar *et al.*, 2006).

Root – Piles (Kumar *et al.*, 2006).

## *BIOPHYTUM SENSITIVUM* (L.) DC. (OXALIDACEAE)

***Syn.:*** *Oxalis sensitiva* L.

***Vern.:*** Lajalu.

***Regional and Other Names :*** *Bengali* - Jhalai; *English* - Sensitive wood-sorrel; *Gujarati* - Jharera; *Hindi* - Lajalu; *Kannada* - Jalapushpam; *Malayalam* - Mukkuti; *Marathi* -Jharera; *Sanskrit* – Lajjaluka, Almbusa; *Tamil* - Mukkutt; *Telugu* – Jalapushpamu.

***Distribution :*** Throughout the hotter parts of India.

***Taxonomic description :*** Erect unbranched annual herbs; leaves long, sensitive, leaflets opposite, sessile, oblong, terminal longest; flowers yellow, in terminal umbels; capsule ellipsoid; seeds ovoid.

***Flowering & Fruiting :*** October – January.

***Medicinal parts :*** Whole plant, leaves, seeds.

***Taste :*** Tasteful.

***Medicinal properties :***

अलम्बुषा खरत्वक् च तथा मेदोगला स्मृता ।
अलम्बुषा लघुः स्वादुः क्रिमिपित्तकफापहः ।। (भा.प्र.)

Alṃbusā khaṛtvak ca tathā medogalā ṣmṛitā
Alṃbusā laghuḥ ṣwaduḥ kṛimi piṭtakaphācahaḥ. (Bha. Pra.)

Anthelmintic, antiseptic, astringent, bitter, cooling, diuretic, stimulant, styptic, tonic.

***Medicinal uses :*** Asthma, chest complaint, cuts, fever, gonorrhoea, tuberculosis, wound.

***Ethno- Medicinal uses :*** Root - Gonorrhoea (Singh *et al.*, 1998).

Leaf - Antifertilizer (Bedi, 1978); wound (Shah *et al.,* 1981; Gupta *et al.,* 1991); antiseptic (Baruah & Sarma, 1984); fever & headache (Kharkongor & Joseph, 1981); rheumatism (Bhatt & Sabnis, 1987).

Whole plant- Wounds, cuts (Kadavul & Jisha, 2006).

## *BOERHAVIA DIFFUSA* L. (NYCTAGINACEAE)

***Syn.:*** *B. repens* L.; *Boervia procumbens* Roxb.

***Vern.:*** Punarnava.

***Regional and Other Names :*** *Bengali*- Punarnava; *English* - Horse puruni spreading hog weed; *Gujarati* - Motosatodo; *Hindi* - Gadhapurna; *Kannada* - Balavadikae; *Malayalam* - Vasu; *Marathi* – Raktavasu; *Sanskrit* – Bhauma, Punarnava; *Tamil* - Kukaratte; *Telugu* – Keerai.

***Distribution :*** Throughout India.

***Taxonomic description :*** Decumbent or diffuse, perennial, extensive herbs. Leaves unequal, ovate-oblong. Flowers rose or pink, in panicles or subcapitate umbels. Fruits clavate, 5-ribbed, covered with stalked glandular hairs.

***Flowering & Fruiting :*** Throughout the year.

***Medicinal parts :*** Fresh whole plant, root, leaves, flowers.

***Taste :*** Bitter.

***Chemical constituents :*** Plant contains Alkaloids – Punarnavan, Sterol, β- Sitosterol, œ- Sitosterol, Elentine, Stearic acid, Palmitic acid, Minerals, potassium nitrate, sodium & sulphate & chloride.

***Medicinal properties :***

पुनर्नवा भवेदुष्णा तिक्ता रूक्षा कफावहा ।
सशोथपांडु हृद्रोगकासोरः क्षतशूलनुत् ।। (ध. नि.)

Punaṛnavā bhavedușṇā tiktā ruksā kaphāpahā
Saśophāpaṇdu hṛdṛogakasoraḥ ksataśūlanuṭ. (Dha. Ni.)

Anti- bacterial, anti- inflammatory, antiviral, astringent, bitter, cooling, diaphoretic, laxative, stimulant, stomachic.

***Medicinal uses :*** Anaemia, asthma, biliousness, blood impurities, bronchitis, cough, epilepsy, eye trouble, heart diseases, inflammation, Jaundice.

***Ethno- Medicinal uses :*** Whole plant - Kidney complaint, gonorrhoea, dropsy (Bhalla *et al.,* 1982); piles, cough (Rajwar G.S., 1983); fever (Bedi, 1978); rheumatism & blood purifier (Janardhanan, 1963); abortifacient (Maheshwari & Singh, 1987); asthma (Sharma *et al.,* 1979); jaundice (Das *et al.*, 1983); dysentery (Maheshwari *et al.*, 1986); stomach trouble (Tripathi *et al.,* 1996; Sudhakar & Chetty, 1998); Inflammation (Singh *et al.*, 1980).

Leaves- urinary problems (Pereira, 2006); spermatorrhoea, leucorrhoea (Prakash & Singh, 2006.); asthma, worm affection, headache, liver and abdominal problem, cough, heart problems (Sharma *et al.,* 2006).

Root - Eczema (Ganeshan *et al.*, 2006); jaundice (Narain & Singh, 2006b; Deokule, 2006); spermatorrhoea (Kumar & Chauhan. 2005); conjuncitivis, blindness (Kumar *et al.,* 2006).

Seed- Jaundice (Jadeja *et al.,* 2006).

## *BOMBAX CEIBA* L. (BOMBACACEAE)

***Syn.:*** *B. malabaricum* DC.; *Salmalia malabarica* (DC.) Schott. & Endl.

***Vern.:*** Semal.

***Regional and Other Names :*** *Bengali* - Shimool; *English* - Silk cotton; *Gujarati* - Shimul; *Hindi* - Semul; *Kannada* - Booruga; *Malayalam* - Poola; *Marathi* - Simalo; *Sanskrit* - Dirghayu, Salmali; *Tamil* - Mullilavau; *Telugu* – Buruga.

***Distribution :*** Throughout the hotter parts of India.

***Taxonomic description :*** Deciduous trees, with prickly trunk and digitately 5 to 7-foliolate leaves. Flowers crimson or yellow, crowded at the end of leafless branches. Capsules woody, 5-valved.

***Flowering & Fruiting :*** February – May.

***Medicinal parts :*** Root, bark, flower, fruit, gum.

***Chemical constituents:***

Seed contains Hexacosanol, tocopherol, terenes.

Bark contains Lupeol, B- sitosterol.

Root contains Triacontanol, B- sitosterol.

Flowers (Petals) contain Essential oil, Hentriacontane, Hentriacontanol, Quercetin, Kaempferol, B- sitosterol.

***Medicinal properties :***

शाल्मलिः शीतला स्वाद्वी रसे पाके रसायनी ।
श्लेष्मला पित्तवातास्त्रेहारिणी रक्तपित्तजित् ।। (भा. प्र.)

Śạlmaliḥ sītlā ṣvādvdī rase pāke rasāyanī
Śleṣhmalā pịttavātāṣṭrahāriṇī raktapịttajịt. (Bha. Pra.)

Root - Astringent, cooling, demulcent, diuretic, sweet, tonic.

Bark - Acrid, astringent, demulcent, diuretic, tonic.

Flowers - Acrid, astringent, bitter, cooling.

Fruits - Alterative, aphrodisiac, cooling, digestive, diuretic, expectorant, stimulant, sweet, tonic.

Gum - Acrid, alterative, aphrodisiac, astringent, demulcent, tonic.

***Medicinal uses*** : Root - Biliousness, Inflammation.

Flower - Blood purifier, leucorrhoea.

Fruits - Leprosy.

Gum - blood diseases, diarrhoea, dysentery, leucorrhoea, monorrhagia.

***Ethno- Medicinal uses :*** Root - Abdominal pain & dysentery (Tarafder, 1983); cough (Jain & Borthakur, 1980); boils (Tarafder, 1984); burn (Pal & Srivastava, 1976); diarrhoea (Rajwar, 1983); promote conception (Jadhav, 2006); spermatorrhoea, gonorrhoea (Prakash & Singh, 2006).

Bark - Pimples & boils (Saxena & Vyas, 1981); dysentery (Malhotra & Moorthy, 1973); bone fracture (Chaudhary *et al.,* 1982); skin diseases (Rajwar, 1983); sexual debility (Ved Prakash & Mehrotra, 1987).

Leaf - Anaemia (Gupta, 1981); rheumatism (Chelvan, 1998).

Gum - Diarrhoea, dysentery (Sharma *et al.,* 1979); asthma (Rajwar, 1983); menorrhoea, leucorrhoea, (Singh & Aswal, 1992; Singh, 1993).

Seed - Small pox (Saxena *et al.,* 1981); liver trouble (Das *et al.,* 2006).

Fruits - Dysentery (Maheshwari & Singh, 1984).

Stem - Gonorrohoea & syphilis (Tiwari *et al.*, 1979).

Flower - Sanke bite (Negi *et al.*, 1985).

## *BUTEA MONOSPERMA* (Lamk.) Taub. (PAPILIONACEAE)

***Syn.:*** - *B. frondosa* Koenig ex Roxb.

***Vern.:*** Palash.

***Regional and Other Names :*** *Bengali* – Palash; *English* - Flame of the forest; *Gujarati* - Khakhra; *Hindi* - Palash; *Kannada* – Mutthuga; *Malayalam* – Chamatha; *Marathi* – Palash; *Sanskrit* - Palasha; *Tamil* - Palasam; *Telugu* – Palasmu.

***Distribution :*** Throughout the greater parts of India.

***Taxonomic description :*** Medium-sized trees, with 3-foliolate large leaves and orange, showy flowers in racemes. Pods grey-silky.

***Flowering & Fruiting :*** March – June.

***Medicinal parts :*** Bark, flowers, seeds, gum.

***Taste :*** Bitter.

***Chemical constituents :*** Plant contains Fats, oils, Mucilage, protein, Resin, Butin, Glucoside, Butein, Isomonospermoside, Coreopsin, Sulphruein, Palastirin, butrin, Isobutrin.

Gum contains Thiamine, Riboflavin, fatty acids, myristi acid, palmitic & stearic acid.

***Medicinal properties :***

पलाशो दीपनो वृष्यः सरोष्णो व्रणगुल्मजित् ।
कषायः कटुकस्तिक्तः स्निग्धो गुदजरोगजित् ।।
भग्नसंधानकृध्दोषग्रहण्यर्शः कृमीन् हरेत्। (भा.प्र.)

Palāśo dīpno vṛṣyaḥ sarasṇo vṛangulmajiṭ
Kasāyaḥ katukaṣtiktaḥ ṣnigdho gudaiarogajiṭ
Bhagnasaṇdhānakṛughdosgrahṇyarśhaḥ kṛumīṇ hareṭ **(Bha.Pra.)**

Bark - Acrid, anthelmintic aphrodisiac, appetiser, bitter, hot, laxative.

Gum - Astringent.

Flower - Acrid, astringent, bitter, hot, sweet.

Fruits - Anthelmintic, aperient, digestive, dry, hot.

Seed - Anthelmintic, aperient, digestive, dry, hot.

***Medicinal uses :*** Root - Elephantiasis, nightblindness.

Bark - Bone fractures, dysentery, hydrocele, piles, tumour, ulcer.

Leaf - Eye diseases

Gum - Cough, dysentery, stomach troubles.

Flower - Burning sensation, eye diseases, gout, leprosy, skin diseases, strangury.

Fruits - Abdominal pain, piles, scorpion sting, skin diseases, tumours, urinary discharges.

Seed - Abdominal pain, piles, scorpion sting, skin diseases, tumours, urinary discharges.

***Ethno- Medicinal uses :*** Root - Dogs bite (Venugopal & Raghunathan, 1982).

Seed - Ascariasis (Jain & De., 1966); bodyache (Jadhav, 2006).

Leaf - Boils (Singh & Maheshwari, 1983); skin diseases (Malhotra & Moorthy, 1973); diarrhoea (Mahanty *et al.,* 1996); ophthalmia (Tripathi *et al.,* 1996); liver problem (Purkyastha & Subhan, 2006); cough & cold (Ganeshan *et al.*, 2006); intestinal worms (Jadeja *et al.,* 2006).

Gum - Antifertilizer (Billore & Audichya, 1978); piles (Singh, 1988); diarrhoea (Sudhakar & Rolla, 1985; Kataria & Gauchan, 2006; Kumar *et al.,* 2006); dysentery (Maheshwari *et al.,* 1986; Singh & Aswal, 1992; Kataria & Gauchan, 2006; Kumar *et al.,* 2006); diabetes (Kumar *et al.,* 2006).

Bark - Bone fracture (Maheshwari & Singh, 1984; Tripathi *et al.,* 1996); wound (Shah *et al.,* 1983; Yesodharan & Sujana, 2006); dysentery (Maheshwari *et al.,* 1986; Tripathi *et al.,* 1996; Narain & Singh, 2006b); diarrhoea (Joshi *et al.,* 1980).

Flower- Piles & diarrhoea (Bhalla *et al.,* 1982); skin diseases (Tripathi *et al.,* 1996); stomach disorders (Lal & Yadav, 1983); pimples (Rajwar, 1983); eye complaint (Shah *et al.,* 1983); heat stroke (Narain & Singh, 2006b); dysentery (Kamble & Pradhan, 1980); antifertilizer (Ramachandran & Nair, 1981).

## *CAESALPINIA BONDUC* (L.) Taub. (CAESALPINIACEAE)

***Syn.:*** *C. bonducella* Fleming.; *C. crista* L.; *C. nuga* Ait. f.; *Guilandina bounducella* L.

***Vern.:*** Kantak karanj.

***Regional and Other Names :*** *Bengali* – Nata; *English* - Fever nut; *Gujarati* - Kakcha; *Hindi* - Katkaranj; *Malayalam* – Kalanji; *Marathi* – Kanchaki; *Sanskrit* – Kantakini, Kantaki karanja; *Tamil* - Kachukkai; *Telugu* – Gutsakai.

***Distribution :*** Throughout India.

***Taxonomic description :*** Armed shrubs, pinnae 6-8 pairs, with a pair of hooked stipular spines; leaflets 6-9 pairs, elliptic-oblong, obtuse. Racemes often panicled; pods shortly stalked, oblong, densely armed on the faces with wiry prickles, 1-2 seeded.

***Flowering & Fruiting :*** June – February.

***Medicinal parts :*** Root, leaves, seeds.

***Taste :*** Very bitter.

***Chemical constituents :*** Seed contains Fatty oils, starch, sucrose, caesalpins, bonducellin, fatty acids, phytosterinin, saponins, phytosterols, amino acids, a bitter glucoside-bonducin.

***Medicinal properties :***

स्त्रंसनं कटुकं पाके लघु वातकफापहम् ।
शोथघ्नमुष्णवीर्यं च पत्रं पूतिकरञ्जम् ।। (सु. सू.)

Ṣraṃsanaṃ katukaṃ pāke laghu vātakaphāpahaṃ
Sothaghnamuṣnavīryaṃ ca patraṃ pūtikaraṇjajaṃ. (Su. Su.)

Leaf - Anthelmintic.

Flower - Bitter.

Fruits - Acrid, anthelmintic, aphrodisiac, astringent.

Seed - Anthelminitic, antiperiodic, styptic.

***Medicinal uses :*** Leaf - Elephantiasis, small pox.

Fruit - Leucorrhoea, piles, urinary discharges, wounds.

Seed - Colic, hydrocele, inflammation, leprosy, malaria, skin diseases.

***Ethno-Medicinal uses :*** Seed - Colic & fever (Shah *et al.,* 1983); rheumatism, skin diseases (Singh & Pandey, 1980); wounds (Saxena & Vyas, 1986); swelling (Rajwar, 1983); malaria (Singh & Pandey, 1980; Purkayastha & Subhan, 2006); stomach troubles (Jadhav, 2006); constipation (Jadeja *et al.,* 2006).

Leaf- Piles (Deokule, 2006).

## *CALOTROPIS GIGANTEA* (L.) Ait. f. (ASCLEPIADACEAE)

***Vern.:*** Safed ankra.

***Regional and Other Names :*** *Bengali* – Akanda; *English* - Giant milk weed; *Gujarati* - Akru; *Hindi* - Madar; *Kannada* – Arka; *Malayalam* – Vellerukku; *Marathi* – Akand; *Sanskrit* – Shwetark, Alarka; *Tamil* - Verukku; *Telugu* – Mandaramu.

***Distribution :*** Throughout India.

***Taxonomic description :*** Shrubs or small trees, with elliptic-oblong or obovate-oblong leaves. Corolla-buds ovoid; corona-lobes hairy, shorter than staminal column.

***Flowering & Fruiting :*** November– June.

***Medicinal parts :*** Root, Leaves, Flowers, Milk.

***Chemical constituents :*** Plant contains Mudarol, akundarol, uscharidin calotropin, frugoside, calotroposides – A to G, calotroplupenyl acetate- A, leupeol acetate – B, gigantursenyl acetate – A, gigantursenyl acetate – B.

***Medicinal properties :***

अलर्ककुसुमं वृष्यं लघु दीपनपाचनम् ।
अरोचक प्रसेकार्शः कासश्वासनिवारणम् ।। (भा. प्र.)

Alaṛkakusumaṃ vṛiṣyaṃ laghu dīpanpāchkaṃ
Arocakah p̣rasekāṛsḥ kāsśwāsniwāṛṇaṃ. (Bha. Pra.)

Plant - Alexipharmic, anthelmintic, Purgative.

Root bark - Diaphoretic.

Flower - Analgesic, anthelmintic, astringent, bitter, sweet.

Milk - Bitter, purgative.

***Medicinal uses :*** Plant - Diseases of the spleen, the liver, the abdomen, Leprosy, leucoderma, piles, tumour, ulcer.

Root bark - Asthma, syphilis.

Flower - Ascites, inflammation, tumour.

Milk - Ascites, diseases of the abdomen, leucoderma, tumour.

***Ethno-Medicinal uses :*** Flower - Cough (Kumar & Chauhan, 2005); asthma (Jadhav, 2006).

Leaf - Bodyache (Rao & Jamir, 1982); Burns (Harja & Baishya, 1981); eye complaint (Sudhakar & Rolla, 1985); Sweeling (Singh & Pandey, 1980); ear pus (Jadhav, 2006); abdominal pain, joint pain, skin diseases (Baruah *et al.,* 2006); chicken pox (Arinathan *et al.,* 2006); rheumatism (Das *et al.,* 2006).

Seed - Antifertility (Tarafder, 1983).

Root - Leprosy (Bhalla *et al.,* 1982); dysentery & elephantiasis (Sudhakar & Rolla, 1985); scorpion bite & snake bite (Jain *et al.,* 1973); leucoderma (Purkayastha & Subhan, 2006); abdominal pain, joint pain, skin diseases (Baruah *et al.,* 2006).

Latex - Wound (Shah, 1984); skin diseases, ringworm (Shah *et al.,* 1983); abortifacient (Tarafder, 1983); toothache (Jain *et al.,* 1973); dog bite (Tiwari *et al.,* 1979); asthma (Saxena, 1986); backache, swelling, jaundice, asthma (Jadhav, 2006); Swelling of joints (Deokule, 2006).

## ***CALOTROPIS PROCERA*** (Ait.) R. Br. (ASCLEPIADACEAE)

***Vern.:*** Ankra.

***Regional and Other Names :*** *Bengali* - Swet-Akanda; *English* - Milk weed; *Gujarati* - Ankru; *Hindi* - Akada; *Malayalam* - Vellerukku; *Marathi* - Mandara; *Sanskrit* -Raktark; *Tamil* - Vellerukku; *Telugu* – Chinnajilledu

***Distribution :*** Throughout India.

***Taxonomic description :*** Shrubs or small trees, with ovate or ovate-oblong leaves. Corolla-buds hemispherical; corona-lobes glabrous, equalling or longer than staminal column.

***Flowering & Fruiting :*** Throughout the year.

***Medicinal parts :*** Root, leaves, flowers, latex.

***Taste :*** Slightly bitter.

***Chemical constituents :*** Plant contains Cardiac steroid glycoside-terpenoids, resin, glycosides, carbohydrates, aluminium, Iron, magnesium, sodium.

***Medicinal properties :***

अर्कद्वयं सरं वातकुष्ठं कंडू विष व्रणान् ।
निहन्ति प्लीहगुल्मार्शः श्लेष्मोदर शकृत्यिक्रिमीन् ।। (भा.प्र.)

Aṛkadyayaṃ saraṃ vātakuṣthaṃ kaṇdū visa vṛaṇān
Nihaṇti plīhguḷmāṛsaḥ śleṣmodar śakṛtyikṛimīṇ. (Bha. Pra.)

Analgesic, anthelmintic, anti-inflammatory, depurative, digestive, emetic, expectorant, laxative, resolvent, stomachic.

***Medicinal uses :*** Asthma, cough, dropsy, eczema, inflammation, piles, rheumatism, stranguary, ulcer.

***Ethno- Medicinal uses :*** Flower - Cholera (Saxena & Vyas, 1981); Cough (Maheshwari & Singh, 1987; Kumar *et al.,* 2006); leucorrhoea (Kumar & Chauhan, 2005).

Wood ash - Asthma (Shah, 1982).

Leaf - Boils, (Rain, 1985; Purkayastha & Subhan, 2006); Skin diseases (Rain, 1985); Swelling & Rheumatism (Bhalla *et al.*, 1982); eczema (Shah, 1982); cold (Maheshwari & Singh, 1987); abdominal pain (Sharma *et al.,* 2006).

Root - Asthma, cough (Singh & Pandey, 1980); elephantiasis (Bhalla *et al.,* 1982); fever (Maheshwari *et al.,* 1986); swelling (Shekhawat & Anand, 1984).

Latex - Leprosy, dropsy, rheumatism (Bhalla *et al.,* 1982); toothache (Maheshwari & Singh, 1984; Sharma *et al.,* 2006); wound (Virjee *et al.,* 1984). Antiseptic (Lad & Yadav, 1983); scorpion sting (Singh & Pandey, 1980); swelling (Jadhav, 2006; Deokule, 2006; Sharma *et al.,* 2006); cough (Kumar *et al.,* 2006); Gout (Jadeja *et al.,* 2006).

Stem - Piles (Mokat & Deokule, 2006).

## *CANNABIS SATIVA* L. (CANNABINACEAE)

***Vern.:*** Bhang.

***Regional and Other Names :*** *Bengali* – Bhang; *English* - True hemp; *Gujarati* - Bhang; *Hindi* - Bhang; *Kannada* – Bangi; *Malayalam* - Kanchavu; *Marathi* – Bhang-Bhanga; *Tamil* - Bhangi; *Telugu* –Ganjayee.

***Distribution :*** Throughout India.

***Taxonomic description :*** Dioecious, aromatic, woody herbs with 3 to 9-foliolate leaves. Flowers white or cream, males in panicels, female's axillary, solitary forming spikes.

***Flowering & Fruiting :*** January- April.

***Medicinal parts :*** Leaves, the part of male flowering twig resin.

***Taste :*** Bitter.

***Chemical constituents :*** Plant contains Cannabinoids, cannabispirans, alkaloids, cannabig lendol, cannabidiolic acid, resin (Cannabinone), gum, sugars, calcium, phosphate, and volatile oil.

***Medicinal properties :***

भंगा कफ हरी तिक्ता ग्राहिणी पाचनी लघुः ।
तीक्ष्णोष्णा पित्तला मोहमदवाग्वह्निवर्द्धिनी ।। (भा.प्र.)

Bhaṇgā kaphaharī tiktā grāhiṇī pācanī laghuḥ
Tīkṣnoṣṇā piṭtalā mohamadavāgvaḥnivardḥanī (Bha. Pra.)

Analgesic, antispasmodic, aphrodisiac astringent, digestive, diuretic, narcotic, stiumulant, stomachic.

***Medicinal uses :*** Leaves - Anthelmintic, cuts, ear trouble, skin diseases, wounds.

Flowers - Diarrhoea, dysnetery, gonorrhoea.

***Ethno- Medicinal uses :*** Leaves - Piles (Lal & Yadav, 1983; Mishra & Billore, 1983; Shah & Joshi, 1971; Singh, 1988); narcotic (Jain & De, 1966); anthelmintic (Shah & Joshi, 1971); bronchitis (Lal & Yadav, 1983); dyspepsia, gonorrhoea (Bhalla *et al.*, 1982); ear complaint, cuts (Shah & Joshi, 1971); eye diseases (Shah, 1982); skin diseases (Rao, 1981; Nautiyal, 1981); pimples, boils, cuts and wounds (Jubilee *et al.*, 2006); diarrhoea, dysentery (Kataria & Gauchan, 2006).

Fruits - Laxative (Rajwar, 1983).

Seeds - Labour pain (Tarafder, 1984).

Bark - Hydrocoel (Sharma *et al.*, 1979).

Root- Gout (Jadeja *et al.*, 2006).

## *CASSIA FISTULA* L. (CAESALPINIACEAE)

***Syn.:*** *C. rhombifolia* Roxb.; *C. excelsa* H. B. & K.

***Vern.:*** Amaltas.

***Regional and Other Names :*** *Bengali* - Bandarlate, Amultas; *English* - Indian laburnum; *Gujarati* - Garmalo; *Hindi* - Amaltas, Bandarlauri; *Kannada* – Kakke; *Malayalam* - Konna, saturangulam; *Marathi* - Bahava, bawa, boya; *Sanskrit* - Rajvriksha, dirdhaphaala, Chaturangula; *Tamil* - Konnai; *Telugu* – Kondrakaya.

***Distribution :*** Throughout India.

***Taxonomic description :*** Deciduous tree; branchlets glabrous; leaves eglandular; leaflets 4-8 pairs,ovate, silvery, pubescent beneath when young; flowers in lax racemes; pods long, cylindrical, nearly straight, smooth, shining, brown-black.

***Flowering & Fruiting :*** March – May.

***Medicinal parts :*** Whole plant.

***Taste :*** Sweet

***Chemical constituents :*** Fruit contains Glycosides, carbohydrates, resin, tannins, magnessium, calcium, potassium, iron & other compounds are fistucacidin, beta-sitosterol, n-tricontanol, leucopelavgonidin, flavonoids, glucosides, gum, and volatile oil.

***Medicinal properties :***

आरावधो गुरूः स्निग्धः शीतलः स्त्रसनो मृदुः ।
ज्वरहृद्रोगपित्तास्त्रवातोदावर्तशूलनुत् ।। (भा. प्र.)

Ārāgvadho guruḥ ṣnigdhaḥ śītalaḥ ṣtransano mṛuduḥ
jvaraḥrogapịttāṣṛtavātodāvaṛtaśūlanuṭ. (Bha. Pra.)

Root – Astringent, cooling, febrifuge, purgative, tonic.

Bark - Anthelmintic, diuretic, emetic, febrifuge, laxative.

Leaves - Antiperiodic, depurative, laxative.

Flower - Acrid, bitter, cooling, demulcent, emollient, expectorant.

Fruit - Abortifacient, anti-inflammatory, antipyretic, cooling, diuretic, emollient, ophthalmic, purgative, sweet.

***Medicinal uses :*** Root - Burning sensation, leprosy, skin disease, syphilis, tuberculous glands.

Bark - Boils, colic, constipation, diabetes, dyspepsia, fever, leprosy, pustules, ringworm.

Leaves - Fever, leprosy, skin disease, ulcers.

Flowers - Burning sensation, bronchitis, cough, pruritus, skin disease.

Fruit - Burning sensation, cardic disorders, colic, fever, gout, inflamation, jaundice, pruritus, rheumatism, Skin disease.

***Ethno- Medicinal uses :*** Root - Leprosy (Barua & Sarma, 1984).

Bark - Blindness (Joshi *et al.,* 1980), dog bite (Tarafdar, 1987), scorpion bite (Roy *et al.,* 1986).

Flowers - Antiseptic (Rajwar, 1983), blood purifier (Srivastava & Verma, 1981), rheumatism (Bhalla *et al.,* 1982), ringworm (Bhalla *et al.,* 1982).

Fruits - Asthma (Tarafdar, 1987), Constipation (Singh & Maheshwari, 1983), Cough (Saxena & Vyas, 1981), diarrhoea (Singh & Maheshwari, 1983; Kataria & Gauchan, 2006), dysentery (Maheshwari & Singh, 1984; Deokule, 2006; Kataria & Gauchan, 2006), pimples (Bhalla *et al.,* 1982), snake bite (Singh & Pandey, 1980), stomachic (Saxena & Vyas, 1981), toothache (Lal & Yadav, 1983); diabetes (Ghosh, 2006).

Seeds - Jaundice (Maheshwari *et al.*, 1981).

Leaf - Leprosy (Purkayastha & Subhan, 2006); skin diseases (Mokat & Deokule, 2006; Kumar *et al.,* 2006).

## *CATHARANTHUS ROSEUS* Don (APOCYNACEAE)

***Syn.:*** *Vinca rosea* L.; *Lochnera rosea* Reichb.

***Vern.:*** Sadabahar.

***Regional and Other Names :*** *Bengali* -Nayantara; *English*-Peri – winkle; *Hindi*-Sadabahar; *Kannada* - Kempukesi; *Malayalam* -Kasithupa; *Marathi* - Sadaphuli; *Sanskrit*-Nityakalyani; *Tamil*-Pottipu; *Telugu* – Billaganneru.

***Distribution :*** Native of America, Very common in Indian Gardens.

***Taxonomic description :*** An erect herb, with simple opposite leaves; white or rosy flowers.

***Flowering & Fruiting :*** Round the year.

***Medicinal parts :*** Whole plant.

***Taste :*** Bitter.

***Chemical constituents :*** Plant contains Indole alkaloids- vincamine; BisIndole alkaloids–vinblastine. Vincristine; Aracidic acid, Behenic acid, Lauric acid, Linolenic acid, oleic acid, palmitic acid, Stearic acid (Seeds).

Stem contains β – Sitosterol, Ursolic acid.

***Medicinal properties :***

नित्यकल्याणिका प्रोक्ता पैतृकी पितृमल्लिका ।
सदापुष्पी मेहहा च उषपुष्पी च गन्धहा ।। (भा.प्र.)

Nịtyakaḷyanikā p̣roktā paiṭriki piṭrumaḷlikā
Sadāpuṣpi mehahā ca usapuṣpi ca gaṇdahā. (Bha. Pra.)

Acrid, astringent, bitter, laxative, stomachic, toxic.

***Medicinal uses :*** Diabetes, haemorrhagic diseases, leucomea, menorrhagea.

***Ethno- Medicinal uses :*** Whole plant - Cancer, diabetes (Baruah & Sarma. 1984).

Root - Cancer (Sudhakar & Rolla, 1985).

Leaf - Menorrhoea (Baruah & Sarma, 1984); swell body (Uniyal & Chauhan, 1973); diabetes (Deokule, 2006; Ghosh, 2006); headache (Das *et al.*, 2006).

## *CAYRATIA CARNOSA* (Wall.ex Wight) Gagnep. (VITACEAE)

***Syn.:*** *Vitis carnosa* L.; *Vitis trifolia* L.

***Vern.:*** Khatkhatumra.

***Regional and Other Names :*** *Bengali* - Amallata; *English* - Fox – grape; *Gujarati* - Khat; *Hindi* - Amalbel; *Kannada* – Heggoli; *Malayalam* – Amarcakkoti; *Marathi* – Amlatbel; *Sanskrit* -Gandirah; *Tamil* - Kattuppirandai; *Telugu* – Kurudinne.

***Distribution :*** Throughout India.

***Taxonomic description :*** Herbaceous climbers, with 3- foliolate leaves and greenish - white flowers in axillary or pseudo, terminal cymes. Berries obovate- globose, dark purple to black when ripe.

***Flowering & Fruiting :*** July – October.

***Medicinal parts :*** Whole plant.

***Taste :*** Acidic.

***Medicinal properties :***

गण्डीरं गण्डुलं पाण्डुपत्रं गण्डीरिका भवेत् ।
चव्या च कीटमारी च पर्विका वृत्तपर्विका ।। (अ.म.)

Gaṇdīraṃ gaṇdulaṃ pāṇdupaṭraṃ gaṇdīriKā bhaveṭ
Cavyā ca kītamārī ca paṛvitā vṛittapaṛvikā. (A.Ma.)

Astringent, diuretic, sour.

***Medicinal uses :*** Cardiac disorders, dropsy, fever, haemorrhoids, neuralgia, splenopathy, strangury, tumors, ulcer, wounds.

***Ethno- Medicinal uses :*** Leaf - Boils (Singh, 1945).

Root - Snake bite (Singh, 1945); Dog bite (Jadhav, 2006).

Whole plant - Piles (Kumar & Chauhan, 2005).

## *CELASTRUS PANICULATUS* Willd. (CELASTRACEAE)

***Syn.:*** *C. montana* Roth.; *C. multiflora* Roxb.

***Vern.:*** Malkangani.

***Regional and Other Names :*** *Bengali* – Malkangli; *English* - Climbing staff plant; *Gujarati* - Malkangana; *Hindi* -Malkangani; *Kannada* – Kariganne; *Malayalam* – Valulavam; *Marathi* –Dhimarbel; *Sanskrit* - Jyotismati; *Tamil* - Valuluvai; *Telugu* – Gundumida.

***Distribution :*** Sub-Himalayan tract, Bombay, S. Gujarat, Central India, Madras Presidency.

***Taxonomic description :*** Unarmed, shrubby climbers, with ovate-elliptic leaves and greenish yellow flowers in terminal panicles. Ovary free from the disc.

***Flowering & Fruiting :*** November– February.

***Medicinal parts :*** Bark, leaves, seeds.

***Taste :*** Bitter.

***Chemical constituents :*** Seed contains Concentrate reddish-brownish, bitter and pungent oil, Resin, Tannin, celastrin and paniculatin.

***Medicinal properties :***

ज्योतिष्मती कटुस्तिक्ता सरा कफसमीरजित् ।
अत्युष्णा वामनी तीक्ष्णा वन्हिबुद्धिस्मृतिप्रदा ।। (भा. प्र.)

jyotiṣmatī katuṣtiktā sarā kaphasamīrajịt
Ạtyuṣnā vāmanī tīksnā vahṇibuddiṣmṛtipṛadā (Bha. Pra.)

Bark - Abortifacient, depurative.

Leaf - Antidote.

Seed - Acrid, anti-Inflammatory, aphrodisiac, appetiser, bitter, digestive, diuretic, emetic, emollicent, expectorant, febrifuge, intellect, laxative, promoting, stimulant, thermogenic.

***Medicinal uses :*** Bark - Brain tonic.

Seed - Abdominal disorders, amenorrhoea, asthma, cardiac debility, dysenorrhoea, fever, inflammation, leucoderma, leprosy, paralysis, pruritus, skin diseases.

***Ethno- Medicinal uses :*** Leaf - Headache (Jain, 1965).

Fruits - Skin diseases & scabies (Tribedi *et al.*, 1982); dysentery (Singh & Maheshwari, 1983).

Root - Snake bite (Maheshwari *et al.*, 1986); sex diseases (Saxena *et al.*, 1981); headache (Jain, 1965).

Bark - Wound (Sharma & Malhotra, 1984); abortifacient (Tarafder, 1983); bronchitis (Joshi *et al.*, 1980); gastric complaint (Goel *et al.*, 1984).

Seed - Cold, cough, fever (Singh & Mahshwari, 1983); wound, scabies (Pal & Srivastava, 1976); skin diseases (Upadhye *et al.*, 1986); backache (Virjee *et al.*, 1984); gout (Goel *et al.*, 1984; Jadeja *et al.*, 2006)); leprosy (Sharma *et al.*, 1979); paralysis (Sebastian & Bhandari, 1984).

Seed oil - Eczema (Maheshwari, 1986); rheumatism (Tribedi *et al.*, 1985); bodyache (Pal & Srivastava, 1976); stomach pain (Srivastava & Varma, 1981).

## *CELOSIA ARGENTEA* L. (AMARANTHACEAE)

***Syn.:*** *C. cristata* L.

***Vern.:*** Shitiwar.

***Regional and Other Names :*** *Bengali* – Swetmurge; *English* - Quail grass; *Gujarati* - Lambdi; *Hindi* - Safed murga; *Kannada* - Anne soppu; *Marathi* – Kurda; *Sanskrit* - Vitunna; *Tamil* -Pannakeerai; *Telugu* – Panchechettu

***Distribution :*** Throughout India.

***Taxonomic description :*** Erect annuals. Leaves linear-lanceolate or ovate-elliptic. Flowers white or pink. Seeds lenticular, black, shining.

***Flowering & Fruiting :*** October – March.

***Medicinal parts :*** Leaves, seeds.

***Taste :*** Acrid.

***Chemical constituents :*** Seed contain Saponin of hyaluronic acid.

***Medicinal properties :***

मूर्वा सरा गुरूः स्वादुतिक्ता पित्तास्त्रमेहनुत् ।
त्रिदोष तृष्णा हृरोगकण्डू कुष्ठजवरापहा ।। (द्रव्यगुण)

Mūrvā sarā guruḥ ṣvādustiktā pịttāṣṭramehnụt
Tridosa ṭṛṣṇā ḥṛdrogakandū kuṣtha jvarāpahā. (Dravyaguna)

Aphrodisiac, astringent, cooling, nutritive, spasmolytic.

***Medicinal uses :*** Seeds - Diarrhoea, dysuria, seminal diseases.

***Ethno- Medicinal uses :*** Seeds - Diarrhoea (Shah *et al.,* 1983); blood purifiers (Singh & Pandey, 1980); sores in mouth (Sharma *et al.*, 1979).

Root- Kidney stone (Deokule, 2006).

Flower- Diarrhoea, dysentery (Kataria & Gauchan, 2006).

## *CENTELLA ASIATICA* (L.) Urban (APIACEAE)

***Syn.:*** *Hydrocotyle asiatica* L.

***Vern.:*** Mandukaparni.

***Regional and Other Names :*** *Bengali* – Tholkhuri; *English* - Indian pennywort; *Gujarati* - Barmi; *Hindi* - Bramha manduki; *Kannada* – Urage; *Malayalam* – Kodangal; *Marathi* – Karivana; *Sanskrit* - Mandukaparni; *Tamil* - Vallarei; *Telugu* – Saraswataku.

***Distribution :*** Throughout India.

***Taxonomic description :*** Creeping herbs, with broadly suborbicular leaves and pink or red flowers in fasciculate umbels. Fruits ovoid.

***Flowering & Fruiting :*** May – November.

***Medicinal parts :*** Whole plant.

***Taste :*** Bitter.

***Chemical constituents :*** Plant contains Alkaloids-Brahmi, Herpestin, Bacosides A & B In Bacoside-A: Arabnosil, glucose, Arabnose, Becogenine, Betulic acid, D-Manitol, Stigmasterol, β-Sitosterol & Tannins.

Leaves contain Glucoside, Volatile oil.

Dry parts contain Centoic acid, centalic acid.

Whole plant also contains Pectic acid, valerin, resin, ascorbic acid.

***Medicinal properties :***

रक्तपित्तरहान्याहुर्हृद्यानी सुलघुनि च ।
कुष्ठमेहज्वर श्वासकासारूचिहरानि च ।। (सु. स.)

Raktapịttahāraṇyāhurhṛdyāni sulaghuni ca
Kuṣthamehajvara ṣvāsakāsāruci harāni ca. (Su. Su.)

Acrid, antileprotic, bitter, cardiotonic, carminative, cooling, diuretic, febrifuge, nervine tonic, stomachic, sweet.

***Medicinal uses :*** Abdominal disorders, amentia, asthma, bronchitis, cardiac debility, epilepsy, fever, hiccough, hoarseness, insomnia, leprosy, strangury.

***Ethno- Medicinal uses :*** Whole plant - Leprosy (Gupta, 1981; Basak, 2006); boils (Kharkongor & Joseph, 1981); jaundice (Hembrom & Goel, 2005); eye infection, stomachache (Baruah *et al.,* 2006); intellegency (Kadavul & Jisha, 2006); skin diseases (Basak, 2006); dysentery (Baruah *et al.,* 2006; Basak, 2006); ulcers, blood purifier (Das *et al.,* 2006).

Fruits - Skin diseases, blood diseases, headache (Singh & Pandey, 1980); Gout (Jadeja *et al.,* 2006).

Leaf - Blood diseases (Nagendra Prasad & Abraham, 1984); stomach trouble, cholera (Rao & Jamir, 1982); gastric diseases & liver complaint (Das *et al.*, 1963); eczema (Purohit *et al.,* 1985); headache (Rao & Neogi, 1980); skin diseases (Sebastian & Bhandari, 1984); syphilis (Karnick *et al.*, 1981; Purkayastha *et al.*, 2006); leprosy (Uniyal & Chauhan, 1973); cough (Karkongor & Joseph, 1981); fever (Pereira, 2006); diarrhoea (Kataria & Gauchan, 2006); dysentery (Kataria & Gauchan, 2006; Rajendran & Rengamani, 2006; Das *et al.,* 2006); constipation (Das *et al.,* 2006).

## ***CENTRATHERUM ANTHELMINTICUM*** (Willd.) Kuntze (ASTERACEAE)

***Syn.:*** *Vernonia anthelminticum* (L.) Willd.; *Serratula anthelmintica* Roxb.

***Vern.:*** Vanjira.

***Regional and Other Names :*** *Bengali* – Somraj; *English* - Purple fleabane; *Gujarati* - Kalijiri; *Hindi* - Somraj; *Kannada* - Kadu – jirage; *Malayalam* – Kalajirakam; *Marathi* – Kalajira; *Sanskrit* – Agnibija, Aranyajirkam; *Tamil* - Kattuchiragam; *Telugu* – Nelavavili.

***Distribution :*** Throughout India.

***Taxonomic description :*** Erect, aromatic, punctate or glandular hairy annuals. Leaves elliptic-obovate or ovate, coarsely serrate. Heads purple. Achenes terete. 10 to 12-ribbed, glandular hairy.

***Flowering & Fruiting :*** August– November.

***Medicinal parts :*** Seeds.

***Taste :*** Bitter.

***Chemical constitucnts :*** Seeds contains β- Sitosterol, Vernolic acid, flavinoids, glycosides, trivernolin, vernodolol, terpineols, sabenene, Isolimonene, caryophyllene, selincnc.

***Medicinal properties :***

अरण्यजीरकं रूक्षं कटुष्णं कटुष्णं दीपनं लघु ।
संग्राही पित्तलं मेध्यं गभशियविशुद्धिकृत् ।।
ज्वरघ्नं पाचनं वृष्यं रूच्यं कफापहम् ।
चक्षुष्यं पवनाध्मानगुल्मच्छर्धतिसार हृत् ।। (भा. प्र.)

Araṇyjirkaṃ rukshaṃ katuṣṇaṃ katuṣṇaṃ dipnaṃ laghu
Saṇgṛāhī piṭtlaṃ meḍhyaṃ gabhśiyviśhuḍdhikṛuṭ
jwraghnaṃ pācnaṃ vṛiṣyaṃ rucyaṃ kaphāpahaṃ
Cakshuṣyaṃ pawanāghnāhuḷmcchṛdyatisār huṭ. (Bha. Pra.)

Seeds - Acrid, anthelmintic, astringent, Sharp taste.

***Medicinal uses :*** Seed - Fever, leucoderma, skin diseases.

***Ethno- Medicinal uses :*** Seed - Leucoderma & skin diseases (Singh & Pandey, 1980); acidity (Chaudhuri & Pal, 1976); asthma (Bedi, 1978); cough (Rajwar, 1983).

Fruits - Paralysis & Swelling (Shah, 1984).

Whole plant - Constipation (Goel & Mudgal, 1988).

## *CHENOPODIUM ALBUM* L. (CHENOPODIACEAE)

***Vern.:*** Bathua.

***Regional and Other Names :*** *Bengali* - Batho shak; *English* - White goosefoot; *Gujarati* - Cheel; *Hindi* - Bathua sag; *Kannada* – Sakothinasoppu; *Malayalam* – Vastuccira; *Marathi* - Chandan bathua; *Sanskrit* - Vastukah; *Tamil* - Parappukeerai; *Telugu* – Pappukoora.

***Distribution :*** Occuring chiefly in cultivated ground.

***Taxonomic description :*** Herbs, with oblong-rhomboid, somewhat fleshy leaves. Flowers pale-green, born in clusters in paniculate, ebracteate spikes; perianth-segements without any swelling on the back. Seeds rugose.

***Flowering & Fruiting :*** November– February.

***Medicinal parts :*** Whole plant.

***Taste :*** Sweet.

***Chemical constituents :*** Plant contains A volatile oil, which contains Ascaridol; saponin, Vitamin- C, Magnessium, Phosphate, Carotene.

***Medicinal properties :***

वास्तुकं वास्तुकन्च स्यात् क्षारपत्रश्च शाकराट् ।
तदेवा तु बृहत्पत्रं रक्तं स्यात् गौडवास्तुकम् ।।
प्रायशो यवमध्ये स्याघवसाकमतः स्मृतम् ।। (भा.प्र.)

Vāṣtukaṃ vāṣtukaṇca ṣyạ̄t ksārapatraśca śākarạ̄t
Tadevạ̄ tu bhṛa̧tpaṭraṃ raktaṃ ṣyạ̄t gaudavạ̄ṣtukaṃ
Pṛāyaśo yavamaḍhye ṣyādyavasākamataḥ ṣmṛtaṃ. (Bha. Pra.)

Acrid, anthelmintic, aphrodisiac, carminative, digestive, diuretic, laxative, Sweet, tonic.

***Medicinal uses :*** Cardiac disorders, dyspepsia, flatulence, general debility, helminthiasis, Peptic ulcer, seminal weakness, strangury.

***Ethno- Medicinal uses :*** Seeds - Skin diseases (Singh & Maheshwari, 1983).

Leaf - Urin complaint (Shah *et al.*, 1983); anthelmintic (Deokule, 2006); anaemia (Kumar *et al.*, 2006).

Leafy twig - General debility & cardiac disorders (Chhetri, 2005).

Root - Jaundice (Shah, 2005); anaemia (Kumar *et al.*, 2006).

## *CHLOROPHYTUM TUBEROSUM* Bak. (LILIACEAE)

***Vern.:*** Safed musli.

***Regional and Other Names :*** *English* - White musale; *Gujarati* - Dhori musli; *Hindi* - Safed mushli; *Sanskrit* –Mushli.

***Distribution :*** E. Himalaya, Assam, Bihar.

***Taxonomic description :*** Annual herbs, with many cylindric root – fibres ending in ellipsoid tubers. Leaves linear – lanceolate. Flowers white, in dense racemes. Capsules obovoid transversely veined.

***Flowering & Fruiting :*** July – August.

***Medicinal parts :*** Root, tubers.

***Taste :*** Tasteless.

***Chemical constituents :*** Root contains Steroids, resin, phenolics, tannins, carbohydrates, calcium, magnessium and potassium.

***Medicinal properties :***

मुशली मधुरा वृष्या वीर्योष्णा बृंहणी गुरूः ।
तिक्ता रसायनी हन्ति गुदजान्यनिलं तथा । (भा.प्र.)

Muśhlī madhurā vṛīṣḥyā viṛyosṇā bṛahiṇī guruḥ
Tiktā rasāyanī haṇti gudjāṇyanilaṃ tathā. (Bha. Pra.)

Root and tuber - Aphrodisiac, cooling, diuretic, tonic.

***Medicinal uses :*** Root and tuber - Diarrrhoea, dysentery, general debility, gonorrhoea, leucorrhoea, seminal debility.

***Ethno-Medicinal uses :*** Tuber - Fever (Roy & Chaturvedi, 1987); leucorrhoea (Saxena & Vyas, 1983); sex tonic (Roy & Chaturvedi, 1986).

## *CICHORIUM INTYBUS* L. (ASTERACEAE)

***Vern.:*** Kasni.

***Regional and Other Names :*** *English* - Chicory; *Gujarati* - Kasani; *Hindi* - Kacani; *Kannada* - Kacani; *Malayalam* - Cikkari; *Sanskrit* - Kasani; *Tamil* - Cikkari; *Telugu* – Cicori.

***Distribution :*** N.- W. India up to 6000 ft.

***Taxonomic description :*** Perennial herbs; radical and lower leaves short petioled, lyrate-pinnatifid, remotely dentate or lobed; upper ones sessile, sagittate with semi-amplexicaul base, entire to coarsely dentate; heads bluish-purple, sessile, 4-10 together in axillary and terminal clusters combined to form panicles; achenes 5-angular, truncate; pappus scales 2-3 seriate.

***Flowering & Fruiting :*** February – May.

***Medicinal parts :*** Whole plant.

***Taste :*** Bitter.

***Chemical constituents :*** Flavonoids, catechol, tannins, 7 bitter sesquiterpene, lactones, sesquiterpene, glucosides, ponticaepoxide, ursolic acid, magnolialide, artesin, oleanolic acid, taraxasterol.

***Medicinal properties :***

कासनी लघुतिक्तोष्णा कफपित्तहरी कटुः ।
यकृद्विकारे हृद्रोगे मूत्रकृच्छे च शस्यते ।। (द्र. विज्ञान.)

Kāsnī laghutiktosṇā kapha pịttaharī katuḥ
Yakṛuḍadvikāre ḥṛdṛoge muṭrakṛucache ca śaṣyate. (Drav. Vij.)

Acrid, anti-inflammatory, alexeteric appetiser, bitter, cardiotonic, depurative, digestive, diuretic, febrifuge, liver tonic, stomachic.

***Medicinal uses :*** Amenorrhoea, asthma, burning sensation, colic, diarrhoea, dysemnorrhoea, dysepepsia, fevers, flatulence, general debility, gout, inflammation, insomnia, jaundice, leprosy, lumbago, ophthalmia, skin diseases, vomiting.

## *CISSAMPELOS PAREIRA* L. (MENISPERMACEAE)

***Syn.:*** *C. hexandra* Roxb.; *C. convolvulacea* Willd.; *C. hernadifolia* Wall.

***Vern.:*** Akauadi.

***Regional and Other Names :*** *Bengali* – Akanadi; *English* - False pareira; *Gujarati* - Venivel; *Hindi* - Karjeuri; *Kannada* – Kodupalli; *Malayalam* - Kattuvalli, Malathangi; *Marathi* – Paharvel; *Sanskrit* - Ambashtha, patha; *Tamil*- Apatta, Panmushtie; *Telugu*–Adavibankateega.

***Description :*** Throughout Tropical and sub-tropical India.

***Taxonomic description :*** Twining, dioecious shrubs, with orbicular or reniform leaves, greenish yellow flowers and subglobose fruits which become red on ripening.

***Flowering & Fruiting :*** July – October.

***Medicinal parts :*** Root, bark.

***Taste :*** Bitter.

***Chemical constituents :*** Root & its bark contains Alkaloids, cissampareine, berberine, cissampeline, pelosine, hayantin, hayatinin, saponin.

***Medicinal properties :***

पाठोष्णा कटुका तीक्ष्णा वातश्लेष्महरी लघुः ।
हन्ति शूलज्वरच्छर्दिकृष्ठातीसारजहृद्रुजः ।।
दाहकण्डुविषश्वासकृमिगुल्मगरव्रणान् ।। (भा. प्र.)

Pāthoṣṇā katukā trīkṣhṇā vātsleṣmharī laghuḥ
Haṇti sūljvarcchrdikṛiṣthātīsārradṛujaḥ
Dāhkaṇduvisasvāskṛumiguḷmgaṛvraṇāṇ. (Bha. Pra.)

Plant - Bitter, hot.

Root - Bitter, stomachic.

***Medicinal uses :*** Plant - Asthma, burning sensation, dysentery, eruptions, fever, heart trouble, itching, piles, skin diseases, ulcers, vomiting.

Root - Conjunctivitis, cough, diarrhoea, dropsy, dyspepsia, scorpion sting, snakebite, stomach pain.

***Ethno- Medicinal uses :*** Leaf - Boils, burns, eye complaint, pimples (Shah & Joshi, 1971); fever, cold (Ramchandran & Nair, 1981); diarrhoea (Singh & Pandey, 1980); dysentery (Pereira, 2006); wound (Shah *et al.*, 1981); stomach diseases (Maheshwari *et al.*, 1980); jaundice (Gogoi & Boissya, 1984); itching (Singh, 1945).

Root - Scorpion bite (Shah *et al.*, 1981); loose motion (Jain, 1965); fever & dog biter (Chaudhuri *et al.*, 1982); diarrhoea (Bhalla *et al.*, 1982; Jadeja *et al.*, 2006); cold (Jain & De, 1966); indigestion & cough (Jain *et al.*, 1973); leucoderma (Purohit *et al.*, 1985); malaria & gastric complaint (Goel *et al.*, 1984); wound (Shah *et al.*, 1983; Gupta, 2006); peptic ulcer (Shah, 1984); colic (Uniyal, 1968); stomach diseases & dysentery (Maheshwari *et al.*, 1986); jaundice (Gogoi & Boissya, 1984); leprosy (Saxena *et al.*, 1981); headache (Trebedi *et al.*, 1985); skin diseases (Kumar *et al.*, 1987); itching (Singh, 1945); piles (Pereira, 2006); stomachache (Gupta, 2006).; snake bite (Shah *et al.*, 1981; Basak, 2006).

## *CISSUS QUADRANGULARIS* L. (VITACEAE)

***Syn.:*** *Vitis quadrangularis* Wall.

***Vern.:*** Hadjor.

***Regional and Other Names :*** *Bengali* - Harjora, harbhanga; *English* - Large granadilla, Bone – setter; *Gujarati* - Chodhari; *Hindi* - Hadjor; *Kannada* – Mangarol; *Malayalam* – Piranta; *Marathi* – Kandawela; *Sanskrit* - Asthisrn-khala; *Tamil* - Pirandai; *Telugu* – Nalleru

***Distribution :*** Throughout India.

***Taxonomic description :*** A large perennial climber. Stems fleshy cactus – like jointed with quadrangular 4 – winged internodes and a tendril at some of the nodes, leaf – ovate or reniform, and are opposed. Flowers greenish - white in shortly peduncled umbellate cymes. Fruits berry, red when ripe.

***Flowering & Fruiting :*** July -August.

***Medicinal parts :*** Whole plant.

***Taste :*** Tasteless (Mucilaginous).

***Chemical constituents :*** Plant contains Carotene, calcium oxalate, vitamin- C, alpha- amyrone and β- sitostosterol.

***Medicinal properties :***

अस्थिसंहारकः प्रोक्तो वातश्लेमहरोअस्थियुक् ।
उष्णः सरः कृमिघ्नश्च दुर्नामघ्नोअक्षिरोगजित् ।। (भा.प्र.)

Aṣthisaṇhārakaḥ p̣rokto vātaṣlesmaharoaṣthiyuk
Uṣṇaḥ saraḥ ḳrmighnaśca duṛnāmaghno aksirogahṛt (Bha.Pra.)

Whole plant - Alterant, anodyne, anthelmintic, aphrodisiac, bitter, carmina-tive, depurative, digestive, haemostatic, laxative, ophthalmic, sour, stomachic, sweet, thermogenic, union - promoting.

***Medicinal uses :*** Whole plant - Chronic ulcers, colic, convulsions, dyspepsia, epilepsy, flatulence, fractures, haemoptysis, haemorrhages, haemorrhoids, helminthiasis, leprosy, ophthalmopathy, otorrhoea, skin diseases, spanomenorrhea, swellings, tumours.

***Ethno- Medicinal uses :*** Stem - Bone fracture (Shah *et al.,* 1983; Jadhav, 2006; Deokule, 2006; Sarma *et al.,* 2006; Basak, 2006); Jiont pain & Swelling (Shah *et al.,* 1983).

Root - Dysentery & Indigestion (Shah *et al.*, 1981); wound & bone fracture (Gupta, 1981).

Whole plant - Bono fracture (Singh & Maheshwari, 1983; Nath *et al.,* 2006).

## *CITRULLUS COLOCYNTHIS* (L.) Schrad. (CUCURBITACEAE)

***Syn.:*** *C. vulgaris* Schard.

***Vern.:*** Indrayan.

***Regional and Other Names :*** *Bengali* - Makai, Indrayan; *English* - Bitter apple; *Gujarati* - Indrak; *Hindi* -Indrayan; *Kannada* – Tumitikayi; *Malayalam* – Peyikummutti; *Marathi* –Indrayan; *Sanskrit* – Visala, Gavadani; *Tamil* - Peykumatti; *Telugu* – Patsakaya.

***Distribution :*** Throughout India.

***Taxonomic description :*** Perennial herbs, with 2 to 3-fid tendrils and deeply 3 to 5-lobed leaves. Flowers greenish-yellow. Fruits globose, mottled with yellowish blotches.

***Flowering & Fruiting :*** November– January.

***Medicinal parts :*** Root, fruits.

***Taste :*** Bitter.

***Chemical constituents :*** Root contains Steroid, œ- Elaterin, Colosynthin, Colosynthothin, saponins.

Seed contains Fixed oil, Phytosterolin, Phytosterol, Saponin, Alkaloid, Glycosides, Tannins, Fatty acids.

Fruits contain Citrulluin, Citrullic acid, Glucose, fructose, Manolic acid.

***Medicinal properties :***

गवादनीद्वयं तिक्तं पाके कटु सरं लघु ।
वीर्योष्णं कामलापित्तकफप्लीहोदरापहम् ।। (भा.प्र.)

Gavāganīdvayaṃ tiktaṃ pāke katu saraṃ laghu
Vīryoṣṇaṃ kāmalāpiṭtakaphaplīhodarāpahaṃ. (Bha. Pra.)

Root - Purgative.

Fruits - Acrid, anthelmintic, antipyretic, bitter, carminative, cooling, purgative.

***Medicinal uses :*** Root- Uteralgia, Mammillitis, Rheumatalgia, Ophthalmia, ascites, jaundice.

Fruits - Ascites, asthma, bronchitis, constipation, dyspepsia, elephantiasis, jaundice, leucoderma, tumours, ulcers.

***Ethno- Medicinal uses :*** Root - Abortifacient (Lal & Lata, 1980); jaundice (Chhetri, 2005; Jadeja *et al.,* 2006); Gout (Jadeja *et al.,* 2006).

Fruits - Stomach trouble (Sen *et al.,* 1985; Arinathan *et al.,* 2006); urinary diseases (Deokule, 2006).

Leaf- Stomach pain (Arinathan *et al.,* 2006).

## *CLEOME GYNANDRA* L. (CLEOMACEAE)

***Syn.:*** *Gynandropsis gynandra* DC.

***Vern.:*** Hulhul.

***Regional and Other Names :*** *Bengali* –Hurhuria; *English*-Spider flower; *Gujarati*-Adiyakharan; *Hindi*-Hulhul; *Kannada* -Narun byalesoppu; *Malayalam* –Kathukatuku; *Marathi* –Satitalvani; *Sanskrit*- Suvarchala; *Tamil*-Naikadugu, Kattkadugu; *Telugu* -Vaminta, velakura.

***Distribution :*** Throughout India.

***Taxonomic description :*** Glandular hairy herbs, with 3 to 5-foliolate leaves. Flowers white, in terminal racemes. Capsules linear-cylindric.

***Flowering & Fruiting :*** June – November.

***Medicinal parts :*** Root, seeds.

***Taste :*** Acrid.

***Chemical constituents :*** Plant contains Essential oil.

Seed contains Cleomin.

***Medicinal properties :***

सुवर्चला हिमा रूक्षा स्वादुपाका सरा गुरूः ।
अम्लपित्तला कटुः क्षारा विष्टम्भकफत्जित् ।। (भा. प्र.)

Suvarchalā himā ruksā ṣwādupākā sarā guruḥ
Aṃlpiṭtlā katuḥ kshārā viṣtaṃbhakaphvaṭjiṭ. (Bha. Pra.)

Root - Hot sharp taste, stomachic.

***Medicinal uses :*** Root - Ascites, earache, fever, spleen enlargement, tumours, ulcers.

***Ethno- Medicinal uses :*** Leaf - Rheumatism, headache, ear complaint (Bhalla *et al.,* 1982); boils (Shah, 1984); headache, rheumatism (Kiruba *et al.,* 2006); earache (Ganesan *et al.,* 2006).

Seed - Cough (Shekhawat & Anand, 1984); headache (Baruah & Sarma, 1984); skin diseases (Bedi, 1978).

Whole plant - Scorpion bites (Nagendra Pradsad & Abraham, 1984); fever (Deokule, 2006).

Root - Fever (Kiruba *et al.,* 2006).

## *CLEOME VISCOSA* L. (CLEOMACEAE)

***Syn.:*** *C. icosandra* L.

***Vern.:*** Hulhur.

***Regional and Other Names :*** *Bengali* – Hurhuria; *English* - Wild mustard; *Gujarati* - Talvani; *Hindi* - Hurhur; *Kannada* – Nayibelu; *Malayalam* – Atunarivela; *Marathi* – Kanphodi; *Sanskrit* -Ajagandha; *Tamil* - Naivelai; *Telugu* – Kukkavaminta.

***Distribution :*** Throughout India.

***Taxonomic description :*** Glandular hariy herbs, with 3 to 5-foliolate leaves. Flowers yellow, in leafy racemes; stamens more than 10. Capsules cylindric, beaked.

***Flowering & Fruiting :*** June – September.

***Medicinal parts :*** Whole plant.

***Taste :*** Bitter.

***Chemical constituents :*** Seed contain Viscosic acid & visosin.

***Medicinal properties :***

अजगन्धा कटुष्णा स्याद्वातगुल्मोदरापहा।
कर्णव्रणार्तिशूलघ्नी कृमिघ्नी च ज्वरापहा ।। (भा. प्र.)

Ajagaṇdhā katuṣaṇā ṣyādvātagulmodarāpahā
Kaṛnavṛaṇāṛttiśūlaghni kṛmighni ca jvarāpahā. (Bha. Pra.)

Plant - Acrid, anthelmintic, sudorific, thermogenic.

Root - Antiscorbutic, stimulant, vermifuge.

Leaves - Digestive.

Seed - Anthelmintic, cardiac stimulant, carminative, constipating, febrifuge.

***Medicinal uses :*** Root - Otaglia.

Seed - Cardiac disorders, diarrhoea, dyspepsia, fever.

***Ethno- Medicinal uses :*** Leaf - Boils (Maheshwari *et al.,* 1980); earache (Shah, 1984; Deokule, 2006); headache (Ramchandran & Nair, 1981); ulcer

(Rajwar, 1983); wound (Bedi, 1978); boils (Kiruba *et al.,* 2006); earache (Kiruba *et al.,* 2006; Kadavul & Jisha, 2006).

Seed - Diarrhoea & fever (Malhotra & Moorthy, 1973; Sharma *et al.,* 2006); headache (Narain & Singh, 2006b); boils & skin diseases (Purohit *et al.,* 1985); intestinal worms (Jadeja *et al.,* 2006); cardiac disorders, dyspepsia (Sharma *et al.,* 2006); earache (Basak, 2006).

fruit – Headache (Sarma *et al.,* 2006).

## *CLERODENDRUM INERME* (L.) Gaertn. (VERBENACEAE)

***Vern.:*** Choti Arni.

***Regional and Other Names :*** *Bengali* – Banjal; *Hindi* - Lanjai, Sangkupi; *Kannada* – Kundali; *Malayalam* –Nirnochi; *Marathi* – Vanajal; *Sanskrit* - Vanajal; *Tamil* - Anjali; *Telugu* – Pisinika.

***Distribution :*** Throughout India.

***Taxonomic description :*** A glabrous shining shrub. Leaves subsessile, variable in size. Flowers white with purple-red filaments in umbelled axillary cymes. Drupes pyriform, subtended by calyx.

***Flowering & Fruiting :*** July – April.

***Medicinal parts :*** Whole plant, Root & Leaves.

***Taste :*** Bitter.

***Chemical constituents :*** Leaves contain Neolignan.

Flower contains Galactose, glucose, fructose, Lysine, arginine, serine, protein, threonine & L– glutamic acid.

Stem & leaves contain Betulin, Neoclerodane, Diterpenoid, cleroinermin, β- sitosterol, clerodermic acid, acacetin, Apigenin.

***Medicinal properties :*** Whole plant - Bitter, blood purifier, fragrant, nutritive.

***Medicinal uses :*** Whole plant - Remitten fever.

Leaf - Scrofula, veneral diseases.

Root - Rheumatism.

***Ethno- Medicinal uses :*** Fruits - Sterilizer (Lal & Lata, 1980).

Leaf- Swelling (Deokule, 2006); rheumatism (Basak, 2006).

## *CLERODENDRUM PHLOMIDIS* L. f. (VERBENACEAE)

***Vern.:*** Arni.

***Regional and Other Names :*** *Gujarati* - Arni; *Hindi* - Arni; *Kannada* - Taggi; *Malayalam* – Tirutali; *Marathi* – Arani; *Sanskrit* - Agnimantha; *Tamil* - Takkari; *Telugu* -Takkillaku.

***Distribution :*** Throughout India.

***Taxonomic description :*** Erect or scandent shrubs, with solid stem and opposite, rhomboid – ovate leaves. Flowers white, in panicles. Fruits obovoid.

***Flowering & Fruiting :*** October – February.

***Medicinal parts :*** Root, Leaves.

***Taste :*** Bitter.

***Chemical constituents :*** Plant contains Pectolinargenin, Scuttelarein, Apigenin, Clerosterol, Hispudilin, Luteolin, Clerodendrin, Cerolic acid, Cerylalcohol, Raffinose, cholesterol, β – Sitosterol.

***Medicinal properties :***

अग्निमन्थः श्वयथुनुद्वीर्योष्णः कफवातहृत् ।
पाण्डुनुत्कटुकस्तिक्तस्तुवरो मधुरोऽग्निदः ।। (भा.प्र.)

Agnimaṇthaḥ svaythunudviṛyosanaḥ kaphavātaṛiṭ
Pāṇdunuṭkatukaṣtiktaṣtuvaro madhuroagnidaḥ. (Bha. Pra.)

Root - Alexipharmic, bitter, heating, laxative, pungent, stomachic.

***Medicinal uses :*** Root - Anemia, constipation, diabetes, dyspepsia, fever, inflammation, piles, swelling.

***Ethno- Medicinal uses :*** Leaf - Headache (Goel *et al.,* 1984); Gout (Jadeja *et al.,* 2006).

Flower - Miscarriages (Jadhav, 2006).

Root- Fever, cough, cold (Sharma *et al.,* 2006).

## *CLERODENDRUM SERRATUM* (L.) Moon. (VERBENACEAE)

***Syn.:*** *C. dentatum* Lam.

***Vern.:*** Bharangi.

***Regional and Other Names :*** *Gujarati* - Bharungi; *Hindi* - Bharangi; *Kannada* – Gantubarangi; *Malayalam* –Cherutekku; *Marathi* – Bharungi; *Sanskrit* – Barangi; *Tamil* - Angaravalli; *Telugu* - Gantu barangi.

***Distribution :*** Throughout India.

***Taxonomic description :*** A shrub, usually herbaceous but occasionally woody, with 4-angular stems. Leaves sessile or subsessile, opposite or ternate. Flowers blue, purple or white, 2-lipped, lips unequal. Drupe broadly obovoid, succulent, dark-purple when ripe.

***Flowering & Fruiting :*** July – November.

***Medicinal parts :*** Roots, leaves.

***Taste :*** Bitter.

***Chemical constituents :*** Roots & leaves contain Glycosides, olianolic acid, querteroic acid, serotonic acid, manitol, saponins, orange coloured resinous substance, starch.

***Medicinal properties :***

भारंगी रूक्षा कटुस्तिक्ता रूच्योष्णा पाचनी लघुः ।
दीपनी तुवरा गुल्मरक्त नुन्नाशयेद् ध्रुवम् ।। (भा. प्र.)

Bhāraṇgī ruksayā katuṣtiktā ruchyoṣnā pācanīlaghuḥ
Dīpanī tuvarā guḷma raktanuṇnāśayeḍ dhṛuvaṃ. (Bha. Pra.)

Root - Acrid, anthelmintic, anti-inflammatory, antispasmodic, bitter, carminative, depurative, digestive, expectorant, febrifuge, stimulant, stomachic, sudorific.

***Medicinal uses :*** Root - Asthma, bronchitis, chronic, colic, consumption, cough, dropsy, dyspepsia, flatulence, fever, helminthiasis, hiccough, inflamma-tion, leprosy, leucoderma, nasitis, skin diseases, tumours, tubercular glands.

***Ethno- Medicinal uses :*** Leaf – Ulcer (Goel *et al.*, 1984); eye diseases (Das *et al.*, 1983);wound (Goel *et al.*, 1984; Kumar *et al.*, 2006).

Root - Fever & malaria (Sudhakar & Rolla, 1985); cough (Jain *et al.*, 1973); bodyache (Tarafder, 1986); snake bite (Maheshwari *et al.*, 1986); rheumatism (Baruah *et al.*, 1987); headache (Kumar *et al.*, 2006).

## *CLITORIA TERNATEA* L. (PAPILIONACEAE)

***Vern.:*** Visnukanta.

***Regional and Other Names :*** *Bengali* – Aparajit; *English* - Butterfly pea; *Gujarati* - Garani; *Hindi* - Aparajita; *Malayalam* – Sankupushpam; *Marathi* – Gokurna; *Sanskrit* - Aparajita; *Tamil* - Kakkattan; *Telugu* – Sankupuspham.

***Distribution :*** Cosmopolitan in the tropics.

***Taxonomic description :*** Twinning herbs, with imparipinnate, 5 to 7-foliate leaves and blue or white flowers. Pods spreading.

***Flowering & Fruiting :*** September – December.

***Medicinal parts :*** Root, Leaves, seeds.

***Taste :*** Bitter.

***Chemical constituents :*** Root bark contains Starch, Tannin, Resin.

Seeds contain Oil, bitter blue coloured resin.

***Medicinal properties :***

अपराजिते कटुमेध्ये शीते कण्ठये सुदृष्टिदे।
कुष्ठमूत्रत्रिदोषामशोथव्रणविषापहे ।।
कषाये कटुके पाके तिक्ते च स्मृतिबुद्धिदे। (भा.प्र.)

Aparājite katu medhyeśīte kaṇthye sudṛṣtide
Kuṣthamutrattidosāmaśothavṛanavisāpahe
Kasāye katuke pāke tikte ca smṛitibuḍdhide. (Bha. Pra.)

Root - Alexeteric, anthelmintic, aphrodisiac, bitter, depurative, diuretic, intellect promoting, laxative, ophthalmic, refrigerant, tonic.

Seed - Cathartic.

***Medicinal uses :*** Root - Amentia, asthma, bronchitis, burning sensation, elephantiasis, fevers, helminthiasis, inflammation, leprosy, leucoderma, Ophthalmopathy, strangury, tubercular glands, ulcers.

Leaves - Eruptions, hepatopathy, Otalgia.

Seed - Visceralgia.

***Ethno- Medicinal uses :*** Root - Abortifacient (Ramachandranan & Nair, 1981); goiter (Saxena & Vyas, 1983); pergative (Malhotra & Moothy, 1973); stomachache (Lakshamanan & Sankaranarayanan, 1988); toothache (Subhangi & Patil, 2006); impotency (Das *et al.,* 2006).

Seed - Laxative (Shah *et al.,* 1983); dropsy (Subhangi & Patil, 2006); stomach trouble (Das *et al.,* 2006).

Leaf - Swelling (Nagendra Prasad & Abraham, 1984); dog bite (Subhangi & Patil, 2006).

Whole plant- Snake bite (Deokule, 2006).

## *COCCINIA CORDIFOLIA* Cogn. (CUCURBITACEAE)

***Syn.:*** *C. grandis* (L.) Voight.; *C. indica* W. & A.; *Cephalandra indica* Naud.

***Vern.:*** Kanduri.

***Regional and Other Names :*** *Bengali* – Telckucha; *English* - Little gourd, Ivy gourd; *Gujarati* - Galludda; *Hindi* - Kanduri; *Kannada* – Tondekayi; *Malayalam* – Kovakai; *Marathi* –Toondale; *Sanskrit* - Bimbi; *Tamil* - Kovakai; *Telugu* – Dondakaya.

***Distribution :*** Throughout India.

***Taxonomic description :*** Slender, dioecious climbers; nodes on the mature stem annulate; leaves entire or digitately 3-5 lobed, cordate; flowers white, usually solitary axillary; fruits oblong-obtuse, green.

***Flowering & Fruiting :*** November- March.

***Medicinal parts :*** Root, leaves, fruits.

***Taste :*** Bitter.

***Chemical constituents :*** Root contain β- Amyrin, Lupeol, β- sitosterol.

Fruits contain Beta-amyrin, lupeol, cucerbitacin, β- sitosterol, taraxerol, lycopene, β- carotene, crypto xanthin.

Seeds contain A fat rich in palmitic, Oleic, Linoleic acids.

***Medicinal properties :***

बिम्बी फलं स्वादु शीतं गुरुपित्तास्त्रवातजित् ।
स्तम्भनं लेखनं रूच्यं विबन्धाध्मानकारकम् ।। (भा.प्र.)

Biṃbī phalaṃ ṣvādu śītaṃ gurupịttāṣṛavātajịt
Ṣtaṃbhanaṃ lekhanaṃ rucyaṃ vibaṇdhādhmānakārakaṃ.
(Bha. Pra.)

Root - Aphrodisiac, cooling.

Leaf - Astringent, bitter, cooling, sweet.

Fruit - Antipyretic, astringent cooling, depurative, expectorant, sweet.

***Medicinal uses :*** Root - Burning sensation, vomiting.

Leaf - Dyspepsia, inflammation, jaundice, leprosy, ulcers.

Fruits - Asthma, burning sensation, bronchitis, cough, diabetes, helminthi-asis, jaundice, leprosy, skin diseases.

***Ethno- Medicinal uses :*** Root - Vomiting & headache (Kumar *et al.*, 1980).

Leaf - Diabetes (Bedi, 1978; Deokule, 2006); earache (Pereira, 2006); eye diseases (Rajendran & Rengamani, 2006).

Fruits - Throat affection (Sharma & Malhotra, 1984); diabetes (Singh & Pandey, 1980); cold & cough (Shah & Gopal, 1986).

## ***COCCULUS HIRSUTUS*** (L.) Diels. (MENISPERMACEAE)

***Syn.:*** *Menispermum hirsutum* L.; *C. villosus* DC.

***Vern.:*** Jaljamni.

***Regional and Other Names :*** *Bengali* – Huyer; *English* - Broom creeper; *Gujarati* - Vevdi; *Hindi* - Jamitikibel, Jaljamni, Patalagarudi; *Kannada* - Dusari balli; *Malayalam* – Patalagarudakkoti; *Marathi* - Vasan vel; *Sanskrit* - Patalagarudah; *Tamil* - Kattukkoti; *Telugu* –Dusaratige.

***Distribution :*** Tropical and sub-tropical India.

***Taxonomic description :*** Scandent, dioecious shrubs, with cordate – lanceolate or subdeltoid leaves. Male flowers minute, yellowish – green, in panicles. Fruits reddish - purple, smooth; Seeds transversely rugose.

***Flowering & Fruiting :*** November – April.

***Medicinal parts :*** Root, Leaves.

***Taste :*** Bitter.

***Chemical constituents :*** Leaves contain Trilobine & Isotrilobine.

Root contains D – Trilobine, DL – Coclaurine.

Stem contains β – Sitosterol, ginnol & mono methyl ether of Inositol.

***Medicinal properties :***

पाठोष्णा कटुकातीक्ष्णा वातश्लेष्महरी लघुः ।
हन्तिशूलज्वरच्छर्दिकुष्ठातिसारहृद्रुजंः ।।
दाहकन्डूविषश्वासकृमिगुल्मगर व्रणान् । (भा.प्र.)

Pāthoṣaṇā katukātīkṣṇā vātaṣieṣmaharī laghuḥ
Haṇtisulajvarachhṛdikuṣthātisārahṛdṛujaḥ
Dāhakaṇdu visaṣvāsa kṛmuguḷmagara vṛanāṇ. (Bha. Pra.)

Root - Acrid, alexeteric, alterant, anodyne, antipyretic, aphrodisiac, bitter, carminative, demulcent, depurative, digestive, diuretic, emollient, expectorant, laxative, tonic.

Leaves - Anodyne, cooling, demulcent, expectorant, mucilaginous.

***Medicinal uses :*** Root - Bronchitis, buring sensation, cephalalgia, colic, cough, dyspepsia, flatulence, fractures, general debility, gout, hypertension, intermittent fevers, leprosy, poisonous bites, pruritus, skin diseases, spermatorrhoea, strangury, tubercular glands.

Leaves - Cough, eczema, gonorrhoea, neuralgia, ophthalmia.

***Ethno- Medicinal uses :*** Leaf - Fever (Shah, 1984); Leucorrhoea (Chaudhari *et al.,* 1975); Cuts, Wound (Johi, 1982); Rheumatism (Shah *et al.,* 1983); Eczema (Sudhakar & Rolla, 1985; Basak, 2006); Conjunctivitis (Audichya *et al.,* 1983); eye ailments (Pereira, 2006); gonorrhoea (Shah, 1984; Deokule, 2006); skin diseases (Shah *et al.,* 1983; Basak, 2006).

Root - Fever & Rheumatism (Shah, 1984); Gout (Goel *et al.,* 1984); leprosy (Basak, 2006).

Stem - Stomach diseases (Shah *et al.,* 1983).

## *COIX LACRYMA - JOBI* L. (POACEAE)

***Vern.:*** Motida.

***Regional and Other Names :*** *Bengali* - Kalo–kunch, Gugur; *English* - Job's tears; *Gujarati* - Kasal; *Hindi* -Sankru; *Kannada* – Manjutti; *Malayalam* – Kattugotampu; *Marathi* – Ranjondhala; *Sanskrit* - Gavedhukah; *Tamil* - Kattukuntumani; *Telugu* - Adavi Guruginja.

***Distribution :*** Throughout India.

***Taxonomic description :*** Robust, spongy, annual grass, often rooting at lower nodes. Leaves cordate at base, with a glabrous membranous ligule. Racemes consisting solitary basal female spikelet and several male spikelets upwards with glume hardly or narrowly winged.

***Flowering & Fruiting :*** September – November.

***Medicinal parts :*** Roots, seeds.

***Taste :*** Bitter.

***Medicinal properties :***

गवेधुका कटुः कार्श्यकरी स्वाद्वी कफापहा ।। (कै. नि.)

Gavedhuka katuḥ kaṛśyakari ṣvadvi kaphaphā. (Kai. ni.)

Seeds - Bitter, cathartic, depurative, diuretic, sweet, tonic.

***Medicinal uses :*** Root - Menstrual disorders, straungury.

Seed - Pectoral diseases, pneumonosis.

***Ethno- Medicinal uses :*** Dysentery, diarrhoea, fever, (Jain & Tarafder, 1970).

Root- Menstrual disorders (Basak, 2006).

## *COLEUS AROMATICUS* Benth. (LAMIACEAE)

***Syn.:*** *C. amboinicus* Lour.

***Vern.:*** Patthar chur.

***Regional and Other Names :*** *Bengali* – Paterchur; *English* - Indian borage; *Hindi* - Patharcur; *Kannada* – Karpurahalli; *Malayalam* – Kannik-kurkka; *Marathi* – Pathurchur; *Sanskrit* – Karpuravalli, Ashmabheda; *Tamil* - Karpuravalli; *Telugu* – Sugandhavalkam.

***Distribution :*** Cultivated throughout India.

***Taxonomic description :*** A small perennial, hairy herb. About 1-3 ft. high with thick and fleshy aromatic leaves; flowers light purple.

***Medicinal parts :*** Leaves.

***Taste :*** Bitter.

***Chemical constituents :*** Leaves contain Diterpene, Methylene – quinone, couleon, cyclobuta tucin.

Root contains Naphtho Pyrone (Rocetin) and Dialdehyde, Avolatile oil carvacrol is also found.

***Medicinal properties :***

अश्मभेदो हिमस्तिक्तः कषायो बस्तिशोधनः ।
भेदनो हन्ति दोषार्शोगुल्मकृच्छ्राश्महृद्रुजः ।
योनिरोगान्प्रमेहांश्च प्लीहशुलव्रणानि च ।। (भा. प्र.)

Aśmbhedo himṣtiktaḥ kasāyo baṣtiśodhnaḥ
Bhedno haṇti dosāṛśọguḷmkṛicchāśmhḍrujaḥ
Yonirogānpṛamehāṇśca plīhśolvṛiṇāni ca. (Bha. Pra.)

Leaf - Acrid, anodyne, anthelmintic, appetiser, aromatic, bitter, carminative, constipating, digestive, diuretic, expectorant, stomachic, thrmogenic.

***Medicinal uses :*** Leaf - Asthma, bronchitis, cholera, colic, cough, diarrhoea, dyspepsia, epilepsy, fever, flatulence, hiccough, otalgia, renal & vesical caliculi.

## *COLOCASIA ESCULENTA* (L.) Schott. (ARACACEAE)

***Syn.:*** *C. antiquorum* Schott.

***Vern.:*** Arvi.

***Regional and Other Names :*** *Bengali* – Kachu; *English* - Cocoyam; *Hindi* - Arvi; *Kannada* – Samangadde; *Malayalam* – Shempu; *Marathi* - Kasalu, alu; *Sanskrit* - Alupam; *Tamil* -Samakilangu; *Telugu* – Chamadump.

***Distribution :*** Wild and cultivated throughout the hotter parts of India.

***Taxonomic description :*** Perennial herbs, with tuberous root – stock of various sizes, shape. Leaves cordate or sagittate at the base, entire, purple-blotched. Spadix with an apical barren appendage, shorter than spathes; spathes pale – yellow.

***Flowering & Fruiting :*** July – September.

***Medicinal parts :*** Leaves, Corms.

***Medicinal properties :***

आलूको बलकृत् स्निग्धः गुर्वीहृत्कफनाशिनी ।
विष्टंभकारिणी तैले तालतोऽतिरूचिप्रदः ।। (भा.प्र.)

Ālūko balakṛt ṣnigdhaḥ guṛvihṛtkaphanāsinī
Viṣtaṃbhakāriṇī taile tālatoatirucipṛadaḥ. (Bha. Pra.)

Leaf - Rubefacient, stimulant, styptic.

Corm - Anodyne, demulcent, laxative.

***Medicinal uses :*** Leaf - Internal haemorrhages, otalgia, otorrhoea.

Corm - Haemorrhoids, somatalgia.

***Ethno- Medicinal uses :*** Corm - Cuts, burns, stings of bees (Rao & Jamir, 1982); Wounds (Saxena, 1986).

## *COMMIPHORA MUKUL* (Hook. ex Stocks) Engl. (BURSERACEAE)

***Syn.:*** *Balsamodendron mukul* Hook. Ex Stocks.; *B. wightii* Arnott.; *B. rouxburghii* Stocks.; *Commiphora wightii* (Arn.) Bhandari.

***Vern.:*** Guggul.

***Regional and Other Names :*** *Bengali* – Gugal; *English* - Indian bdellium tree; *Gujarati* - Gugali; *Hindi* - Guggul; *Kannada* – Guggulu; *Malayalam* – Gulgulu; *Marathi* – Guggala; *Sanskrit* - Gugguluh; *Tamil* - Gukkulu; *Telugu* – Guggulu.

***Distribution :*** Bellary, Mysore, Kathiawar, Rajputana Desert.

***Taxonomic description :*** A much branched shrub or moderate tree with silvery, paper like bark peelings; branches knotty and crooked, usually ending in sharp spines; leaves 1-3 foliolate, subsessile, rhomboid-ovate, serrate; flowers brownish-red in fascicles of 2-3; drupe ovoid, red when ripe.

***Flowering & Fruiting :*** March – April.

***Medicinal parts :*** Resinous gum.

***Taste :*** Bitter.

***Chemical constituents :*** Gum-resin contains Sesamin, steroids, essential oil, guggul, sterol, gugguls steron, myrcene, dimercene, polymercene and sugars.

Flowers contain Myrcyalcohol, β- Sito sterol, amino acids, sugars, œ-camphorene, cembrene, allysembrol, flavonoids, & ellagic acid.

Seed oil contains Lenoleic acid, oleic acid, palmitic acid, stearic acid.

***Medicinal properties :***

गुग्गुलुः कटुतिक्तोष्णः कफमारूतकासजित् ।
कृमिवातोदरप्लीहशोफाशोघ्नो रसायनः ।। (रा. नि.)

Gugguluḥ katutiktoṣṇaḥ kaphamārutakāsajịt
Kṛmivātodarplīhaśophārśoghno rasāyanaḥ. (Ra. Ni.)

Acrid, alterative, anodyne, anthelmintic, anti-inflammatory, antiseptic, antispasmodic aphrodisiac, aromatic, astringent, bitter, depurative, digestive, diuretic, expectorant, rejuvenating, stimulant, thermogenic.

***Medicinal uses :*** Amenorrhea, anaemia, asthma, bronchitis, cardiac disorders, cough, diabetes, dysmenorrhoea, dyspepsia, epilepsy, facial paralysis, fever, gout, helminthiasis, leprosy, leucoderma, sciatica, skin diseases, spermatorrhea, ulcers, urinary calculus, wounds.

***Ethno- Medicinal uses :*** Resin - Asthma, boils, headache (Shah, 1982); dysentery (Jadeja *et al.,* 2006).

Bark - Gum trouble & ulcer (Shekhawat & Anand, 1984).

Leaves- Gout (Jadeja *et al.,* 2006).

## *CONVOLVULUS PLURICAULIS* Choisy. (COVOLVULACEAE)

***Vern.:*** Shankpushpi.

***Regional and Other Names :*** *Hindi* - Shankpushpi; *Gujarati* – Shankhavali; *Marathi*- Sankhvel; *Sanskrit* – Shankpushpi.

***Distribution :*** Punjab plains and eastwards to Bihar and Chota Nagpur.

***Taxonomic description :*** A prostrate or suberect, spreading, hairy, perennial herb with a woody rootstock. Leaves linear-lanceolate; hairy. Flowers white or pale-rose, solitary or in pairs, sessile or subsessile in the leaf axils. Capsule oblong-globose.

***Flowering & Fruiting :*** May-December.

***Medicinal parts :*** Whole plant, root, leaves, flowers.

***Taste :*** Bitter.

***Chemical constituents:*** Plant contains Alkaloid shankhpushpin, essential oil, methyl pentos, scamonolic acid and palmintic acid, phytosterol, ipuranol, sucrose, reducing sugars, starch and olein.

***Medicinal properties :***

शंखपुष्पी सरामेध्या वृष्यामानसरोगहृत् ।
रसायनी कषायोष्णा स्मृति कान्तिबलाग्निदा ।। (भा. प्र.)

Śhankhapuṣhpī sarāmedhyā vṛiṣyāmānasaroghṛiṭ
Rasāyanī kasāyoṣṇā ṣmṛitikāṇtibalāgnidā. (Bha. Pra.)

Anthelmintic, brain tonic, laxative, spasmolytic, stomachic.

***Medicinal uses :*** Effective on brain and heart, effective on thyroid high secretion; increases memory, positive effect on blood pressure, reduces blood pressure and tension, useful in epilepsy, insomnia & mental diseaes.

## *CORDIA DICHOTOMA* Forst. f. (BORAGINACEAE)

***Syn.:*** *C. obliqua* Willd.; *C. myxa* Roxb. non L.; *C. latifolia* Roxb.

***Vern.:*** Gunda.

***Regional and Other Names :*** *Bengali* – Bahubara; *English* - Sebesten plum; *Gujarati* - Gundomoto; *Hindi* -Lasura; *Kannada* – Cikkacalli; *Malayalam* – Naruvari; *Marathi* – Bhokar; *Sanskrit* – Bahuvara, Slesmatakah; *Tamil* - Naruvilli; *Telugu* – Cinnanakkeru.

***Distribution :*** Throughout India.

***Taxonomic description :*** Deciduous trees, with 3 to 5-nerved, elliptic-ovate or obovate leaves longer than broad. Flowers white, in dichotomous cymes. Fruits globose, yellow to red when ripe.

***Flowering & Fruiting :*** March- July.

***Medicinal parts :*** Bark, leaf, fruits.

***Taste :*** Bitter.

***Chemical constituents :*** Bark contains Tannins, allantonin, B-sitosterol, a flavanone rhamnopyranoside.

Fruits contain Mucilage & sugars.

Seed oil contain B-sitosterol.

Seeds contain flavonolglycosides, taxifolin-3- rhamnoside.

***Medicinal properties :***

बहुवारस्तु शीतः स्यादुद्दालो बहुवारकः ।
शेलुः श्लेष्मातकश्चापि पिच्छिलो भूतवृक्षकः ।। (भा.प्र.)

Bahuvāraṣtu śitaḥ ṣyāduddālo bahuvārakaḥ
Śeluḥ śleṣmātakascāpi picchilo bhūtavṛksakaḥ. (Bha. Pra.)

Bark-Acrid, anthelmintic, astringent, bitter, constipating, cooling, depurative.

Leaf - Aphrodisiac.

Fruits- Anthelmintic, aphrodisiac, cooling, depurative, diuretic, emollient, expectorant, febrifuge, purgative, sweet.

***Medicinal uses :*** Bark - Burning sensation, diarrhoea, dyspepsia, fever, helminthiasis, leprosy, skin diseses.

Leaf - Gonorrhoea.

Fruits - Bronchitis, burning sensation, cough, fever, leprosy, pectoral diseases, ringworm, skin diseases, ulcers.

***Ethno- Medicinal uses :*** Fruits - Chest Inflammation, lung diseases (Bhalla *et al.*, 1982); stomach diseases (Sudhakar & Rolla, 1985); constipation (Roy & Chaturvedi, 1986); urinary complaints (Deokule, 2006).

Leaf - Chest Infection, lung diseases, colic, stomach diseases (Sebastian & Bhandari, 1984); cold & cough (Sexena & Vyas, 1983); headache (Das *et al.*, 1983); fever (Singh, & Maheshwari, 1983); jaundice (Hemadri & Rao, 1984).

Bark- toothache (Aditya *et al.*, 2006).

## *COSTUS SPECIOSUS* (Koenig ex Retz.) J.E. Smith (ZINGIBERACEAE)

***Syn.:*** *Banksea speciosa* Koenig.

***Vern.:*** Kevuk kand.

***Regional and Other Names :*** *Bengali* - Kust, Keu; *Gujarati* - Pokaramula; *Hindi* - Kust, Keu; *Kannada* – Chengalvakoshtu; *Malayalam* - Anappu, Pushkaramulam; *Marathi* – Pushkarmula; *Sanskrit* -Kebuka; *Tamil* - Kostam; *Telugu* - Kavukinna, Chengalvakoshtu.

***Distribution :*** Throughout India.

***Taxonomic description :*** An erect, herbaceous plant; leaves subsessile, arranged spirally, oblong or oblanceolate, glabrous above, thinly silky beneath; flowers white, large, many, in terminal dense heads; bracts 2-3 cm long, ovate, brownish-red or bright red; capsule subglobosely 3-gonous; seeds black with white aril.

***Flowering & Fruiting :*** July – September.

***Medicinal parts :*** Rhizomes.

***Taste :*** Bitter.

***Chemical constituents :*** Rhizomes contain Alkaloids, Tigogenin, Diosgenine, Starch, β-Sterol, glucoside.

Leaves contain Alpha-Amyrin stearate, β- Amyrin and Lupeol palmitates.

***Medicinal properties :***

केबुकं कटुकं पाके तिक्तं ग्राहि हिमं लघुः।
दीपनं पाचनं हृद्यं कफपित्तज्वरापहम् ।।
कुष्ठाकासप्रमेहास्त्रनाशनं वातलं कटु। (भा. प्र.)

Kebukaṃ katukaṃ pāke tiktaṃ grāhi himaṃ laghuḥ
Dīpnaṃ pācnaṃ hrdyaṃ kaphapịttajvarāpahaṃ
Kuṣthakāsap̣ramehāṣ̣ranāsanaṃ vātalaṃ katu. (Bha. Pra.)

Rhizome - Acrid, anthelmintic, aphrodisiac, astringent, bitter, cooling, depurative, expectorant, febrifuge, purgative.

***Medicinal uses :*** Rhizome - Anaemia, asthma, bronchitis, burning sensation, constipation, fever, flatulence, helminthiasis, hiccough, inflammation, leprosy, skin diseases.

***Ethno- Medicinal uses :*** Leaf - Fever (Krishna & & Singh, 1987).

Whole plant - Ear complaint (Joshi *et al.*, 1980).

Root - Bodyache & fever (Jain & De, 1966); cough, piles (Nath *et al.*, 2006); Gout (Jadeja *et al.*, 2006).

Rhizome - Constipation, dysentery (Jain, 1965); ear complaint (Saxena *et al.*, 1981); jaundice (Rao, 1981); leprosy (Rao, 1981; Purkayastha & Subhan, 2006).

## *CRATAEVA MAGNA* (Lour.) DC. (CAPPARACEAE)

***Syn.:*** *C. nurvala* Buch.- Ham.; *C. religiosa* Hook. f. & Thomas non frost.f.

***Vern.:*** Barna.

***Regional and Other Names :*** *Bengali* – Barun; *English* - Caper tree, garlic pear; *Gujarati* - Varno; *Hindi* - Barna; *Malayalam* – Nirmathalan; *Marathi* – Wayawarna; *Sanskrit* -Varuna; *Tamil* - Mavalingam; *Telugu* – Vulemeri.

***Distribution :*** Almost all over India.

***Taxonomic description :*** Deciduous trees, with ash-coloured bark. Leaves petiolate; leaflets obilquely falcate, acuminate. Flowers creamish-yellow when fresh. Fruits subglobose or ovate-oblong, greyish when dry, with dry flat papillae.

***Flowering & Fruiting :*** Jan. – June.

***Medicinal parts :*** Bark, leaves.

***Taste :*** Bitter.

***Chemical constituents :*** Bark contains Tannin, saponin.

Root bark contains Leupeol, stearic acid, lauric acid, linoleic acid, oleic acid, triter penoids.

***Medicinal properties :***

वरूणः पित्तलो भेदी श्लेष्मकृच्छाममारूतान् ।
निहन्ति गुल्मवातास्त्रकृमींश्चोष्णोऽग्निदीपनः ।।
कषायो मधुरस्तिक्तः कटुको रूक्षको लघुः । (भा. प्र.)

Varunaḥ piṭtalo bhedī sleṣmakṛcchāmamārutāṇ
Nihaṇti gulmavātāṣrakṛmīṣcoṣṇoagni dīpanaḥ
Kasāyo madhuraṣtiktaḥ katuko ruksako laghuḥ. (Bha. Pra.)

Bark and leaves - Acrid, anthelmintic, astringent, bitter, carminative, depurative, digestive, diuretic, expectorant, laxative, stimulant, stomachic, thermogenic.

***Medicinal uses :*** Bark and leaves - Asthma, bronchitis, colic, cough, fever, flatulence, helminthiasis, inflammation, pruritus, renal calculus, skin diseases, tubercular gland.

***Ethno-Medicinal uses :*** Bark - Fever & stomach pain (Nagaendra Pradsad & Abraham, 1984); urine diseases (Jain & Borthakur, 1980); abortifacient (Tarafder, 1983); wound (Jain, 1965).

Leaf- Dysentery, dyspepsia (Jadeja *et al.,* 2006).

## *CRINUM ASIATICUM* L. (AMARYLLIDACEAE)

***Syn.*** *C. toxicarium* Roxb.; *C. bracteatum* Willd.

***Vern.:*** Sudarshan.

***Regional and Other Names :*** *Bengali* - Baro kanur; *English* - Poison bulb; *Gujarati* - Nagdamni; *Hindi* - Cindar; *Kannada* –Vishamungali; *Malayalam* – Polathali; *Marathi* – Nagdouna; *Sanskrit* -Nagadamni; *Tamil* - Vishamungil; *Telugu* – Vishamungali.

***Distribution :*** Throughout India.

***Taxonomic description :*** Bulb narrowed into a neck which is clothed with old leaf –sheaths. Leaves linear – lanceolate, flat, narrow, with a sheathing base, bright – green. Scape long. Flowers white, fragrant at night; bracts long; pedicels shorter than bracts; perianth – tube greenish- white.

***Flowering & Fruiting :*** April – May.

***Medicinal parts :*** Bulb.

***Taste :*** Bitter.

***Chemical constituents :***

Fruits contain Alkaloid-Hamayne

Bark contains Sterol, triterpines

Root contains Alkaloid- Narcissine (Lycorine) & Crinamine.

***Medicinal properties :***

विज्नेया नागदमनी बलामोटा विषापहा ।
नागपुष्पी नागपत्री महायोगेश्वरीति च ।। (भा.प्र.)

Vijneyā nāgadamanī balāmotā visāpahā
Nāgapuṣpī nāgapaṭrī mahāyogesvarīti ca. (Bha. Pra.)

Bulb - Acrid, anthelmintic, anti– inflammatory, antipyretic, aphrodisiac, bitter, carminative, diaphoretic, diuretic, emetic, expectorant, laxative, thermogenic.

***Medicinal uses :*** Bulb - Bronchitis, cough, dysopia, dyspepsia, fevers, flatulence, general debility, gonorrhoea, helminthiasis, inflammation, tumours, wounds.

***Ethno- Medicinal uses :***

Leaves - Skin trouble (Kiruba *et al.,* 2006).

## *CROTALARIA RETUSA* L. (PAPILIONACEAE)

***Vern.:*** Ghunghunia.

***Regional and Other Names :*** *Bengali* – Atasi; *English* - Rattlewort; *Gujarati* - Ghughra; *Hindi* - Ghunghunia; *Kannada* – Guluguluppahalli; *Malayalam* – Kilukiluppa; *Marathi* – Ghagri; *Sanskrit* - Sanapuspi; *Tamil* - Kilukilluppai; *Telugu* – Bottigilligichcha.

***Distribution :*** Throughout India.

***Taxonomic description :*** Erect undershrubs; branches striate, appressedly white pubescent; leaves obovate, oblanceolate, and silky pubescent beneath; flowers in 15-30 flowered terminal racemes; pod exserted glabrous, stalked, linear-oblong; seeds 15-30.

***Flowering & Fruiting :*** September– November.

***Medicinal parts :*** Whole plant.

***Taste :*** Bitter.

***Medicinal properties :***

शणपुष्पी स्मृता घण्टारवा शणसमाकृतिः । (भा. प्र.)

Śaṇapuspi ṣmṛtā ghaṇṭāravā śaṇasamākṛtiḥ. (Bha. Pra.)

Plant - Astringent, bitter, emetic, expectorant.

Seed - Depurative, digestive, febrifuge.

***Medicinal uses :*** Plant - Cardiac disorders, dyspepsia, fever.

Leaf - Diarrhoea, scabies.

Seed - Fever, flatulence, leprosy, skin diseases.

## *CURCULIGO ORCHIOIDES* Gaertn. (AMARYLLIDACEAE)

***Syn.:*** *C. malabarica* Wight.; *C. brevifolia* Dryland.; *Hypoxis orchiodies* Kurz.; *Carcinia malabarica* Desv.

***Vern.:*** Kali musli.

***Regional and Other Names :*** *Bengali* – Talamuli; *English* - Black musali; *Gujarati* - Kalmusli; *Hindi* - Mushali; *Kannada* – Nelatatigadde; *Malayalam* – Nelappanakizhanna; *Marathi* – Kalimusali; *Sanskrit* – Musali, Talamuli; *Tamil* - Nilapanni; *Telugu* – Nelatadi.

***Distribution :*** Bengal, Assam, W. Peninsula.

***Taxonomic description :*** Perennial herbs, with thick tuberous root – stock and without aerial stem. Leaves plicate, linear - lanceolate. Flowers yellow 2 to 3 together on axillary peduncles hidden among leaf-sheaths.

***Flowering & Fruiting :*** June – September.

***Medicinal parts :*** Root-stock.

***Taste :*** Bitter.

***Chemical constituents :*** Root stock contains Steroids–Terpenoids, Phenolics, Tannins, Glycosides, Starch, Iron, Calcium, Magnessium, Sodium, Stranthium, Resin, gum, Calcium Oxalate.

***Medicinal properties :***

तालमूली तु विदवदर्भिर्मुशली परिकिर्त्रिता । (भा. प्र.)

Tālamūlī tu vidvadhiṛmusalī parikiṛttitā. (Bha. Pra.)

Alterative, antipyretic, aphrodisiac, appetiser, bitter, carminative cooling, depurative, diuretic, emollient, sweet.

***Medicinal uses :*** Asthma, bronchitis, colic, diarrhoea, dyspepsia, gonorrhoea, jaundice, leucorrhoea, lumbago, pruritus, skin diseases, vomiting, wounds.

***Ethno- Medicinal uses :*** Whole plant - Cuts & wound (Jain *et al.,* 1973).

Rhizome - Asthma, skin diseases (Nagendra Prasad & Abraham, 1984); blindness, cuts, wounds (Jain *et al.,* 1973); Itching (Janardhanan, 1963); cuts (Jain, 1963); wounds (Malhotra, & Moorthy, 1973); piles (Karnick *et al.*, 1981; Das *et al.,* 2006); kidney trouble & bronchitis (Chhetri, 2005); diarrhoea, leucorrhoea, jaundice (Das *et al.,* 2006).

Root - Filarial, fever (Jain & De, 1966); abortifacient (Tarafder, 1983); cuts, cough, ear complaint (Jain *et al.*, 1973); dysentery (Purohit *et al.*, 1985); epilepsy (Tarafder & Rai chaudhuri,1981); scorpion bite (Saxena & Vyas, 1983); snake bite (Saxena & Dutta, 1975); leucorrhoea (Tarafder & Rai Chaudhuri,1981).

Seed- Gout (Jadeja *et al.,* 2006).

## *CURCUMA AMADA* Roxb. (ZINGIBERACEAE)

***Vern.:*** Ambi haldi.

***Regional and Other Names :*** *Bengali* – Amada; *English* - Mango ginger; *Hindi* - Ama-haldi; *Kannada* – Ambarasini; *Malayalam* – Mangainji; *Marathi* - Ambe haldi; *Sanskrit* - Amragandhiharidra, Karpurdharidra; *Tamil* - Mainji; *Telugu* - Mamide allam.

***Distribution :*** Bengal, Malaya Peninsula, W. Peninsula.

***Taxonomic description :*** Stemless herbs, with ellipsoid, sessile tubers with the smell of unripe mango. Leaves oblong-lanceolate. Flowers white or pale -yellow, in spikes appearing with the leaves.

***Medicinal parts :*** Rhizomes.

***Taste :*** Bitter & Sweet.

***Chemical constituents :*** Rhizome contains Essential oil (Pinene, ocimene, safrol), Resin, sugars, Gum, Starch, Albuminoids, crude fibres, organic acid, Phytosterol, zulenogeine, Camphor, curcumene, Turmerone, Volatile oil.

***Medicinal properties :***

आम्रगन्धिहरिद्रा या सा शीता बातला मता ।
पित्ताह्रन्मधुरा तिक्ता सर्वकण्डू विनाशिनी ।। (भा.प्र.)

Āmṛagaṇdhihariḍra yā sā śitā vātalā matā
Pịttāhṛnmadhurā tiktā saṛvakaṇdū vināśini. (Bha. Pra.)

Anti – inflammatiory, anti – pyretic. aphrodisiac, appetizer, aromatic, bitter, carminative, cooling, digestive, diuretic, expectorant, febrifuge, laxative, sour, stomachic, sweet.

***Medicinal uses :*** Bronchitis, colic, constipation, cough, dyspepsia, fever, flatuance, gout, hiccough, inflammation, pruritus, skin diseases, ulcers, wounds.

***Ethno- Medicinal uses :*** Rhizome- Sprains (Deokule, 2006).

## *CURCUMA LONGA* L. (ZINGIBERACEAE)

***Syn.:*** *C. domestica* Valeton.; *C. rotunda* L.; *Amomum curcuma* Jacq.

***Vern.:*** Haldi.

***Regional and Other Names :*** *Bengali* – Halud; *English* - Turmeric; *Hindi* - Haldi; *Kannada* – Arisina; *Malayalam* – Mannal; *Marathi* – Halad; *Sanskrit* - Haridra; *Tamil* - Mancal; *Telugu* – Pasupu.

***Distribution :*** Cultivated throughout the India.

***Taxonomic description :*** Rhizomes large, yielding turmeric. Leaves oblong-elliptic. Flowers pale-yellow.

***Medicinal parts :*** Rhizomes.

***Taste :*** Slightly bitter.

***Chemical constituents :*** Rhizome contains Active curcumin compound; essential oils-Termeric oil, Terpenoids, A volatile oil: it contains curcumen (Terpene), starch, Albumnoids.

***Medicinal properties :***

हरिद्रा कटुका तिक्ता रूक्षोष्णा कफपित्तनुत् ।
वर्ण्या त्वग्दोषमेहास्त्रशोथपाण्डुव्रणापदा ।। (भा.प्र.)

Haridṛā katukātiktā ruksosṇā kaphapiṭtanuṭ
Vaṛṇyā .tvagdosamehāṣṛasothapāṇduvṛaṇāpahā. (Bha. Pra.)

Rhizome - Acrid, alterative, anodyne, anthelmintic, anti – inflammatory, antiperiodic, antiseptic, appetiser, bitter, carminative, depurative, diuretic, emollient, expectorant, febrifuge, laxative, stimulant, stomachic, thermogenic.

***Medicinal uses :*** Rhizome - Amenorrhoea, anaemia, asthma, bronchitis, colic, conjunctivitis, constipation, cough, diabetes, dropsy, dyspepsia, elephantiasis, fever, flatulence, gonorrohoea, haemorrhoges, helminthiasis, hiccough, hysteria, inflammation, jaundice, leprosy, pruritus, ringworm, skin diseases, ulcers, wounds.

***Ethno- Medicinal uses :*** Rhizome - Anthelmintic (Banerjee, 1977); cough (Shah & Joshi, 1971); headache (Lal & Yadav, 1983); insect stings

(Shah & Joshi, 1971); swell body (Shah & Joshi, 1971; Rajendran & Rengamani, 2006); wounds (Shah & Joshi, 1971; Bhattacahrjee *et al.,* 1980; Tarafder, 1984; Deokule, 2006); leprosy (Purohit *et al.,* 1985); migraine (Tiwari *et al.,* 1979); pimples (Prohit *et al.,* 1985); bodyache (Lal & Yadav, 1983); cuts & wounds (Purkyastha & Subhan, 2006); worm complaints (Udayan *et al.,* 2006; Gupta, 2006); snake bite (Udayan *et al.,* 2006 b); wounds (Deokule, 2006); skin diseases, swelling (Baruah *et al.,* 2006); diarrhoea, jaundice (Jadeja *et al.,* 2006); Hepatitis (Dutta Choudhury & Shil, 2006); dysentery, stomach trouble (Gupta, 2006); Gout (Jadeja *et al.,* 2006); constipation (Das *et al.,* 2006).

## *CUSCUTA REFLEXA* Roxb. (CONVOLVULACEAE)

***Vern.:*** Amarbel.

***Regional and Other Names :*** *Bengali* – Swarnalata; *English* - Dodder; *Gujarati* - Amarabel; *Hindi* - Amarbel; *Kannada* – Amaraballi; *Malayalam* – Akasavalli; *Marathi* – Amarvel; *Sanskrit* -Akasavalli; *Tamil* - Kadiyagundal; *Telugu* – Savarapukada.

***Distribution :*** Throughout India.

***Taxonomic description :*** Leafless stem – parasites, with greenish – yellow stem. Flowers whitish, solitary or in clusters arranged in racemes; style 1; stigmas 2.

***Flowering & Fruiting :*** December- January.

***Medicinal parts :*** Whole plant.

***Taste :*** Slightly Bitter.

***Chemical constituents :*** Plant contains Alkaloids, proteins, flavonoids, resin, tannins, glycosides, carbohydrates, cuscutin, cuscatalin, reducing sugars.

Seeds contain Fixed oil, Amarbelin (Pigment.), bitter resin, aluminum, calcium, sodium, potassium & reducing sugars.

***Medicinal properties :***

खवल्ली ग्राहिणी तिक्ता पिच्छिलाक्ष्यामयापहा।
तुवराग्निकरी हृद्या पित्तश्लेष्मामनाशिनी ।। (भा.प्र.)

Khavallī grāhinī tiktā picchalāksyāmyāpahā
Tuvarāgnikarī hridyā pittaśleshmmāmnāśhinī. (Bha. Pra.)

Whole plant - Anthelmintic, aphrodisiac, astringent, bitter, carminative, diuretic, expectorant, sweet.

***Medicinal uses :*** Whole plant - Bronchitis, cough, fever, Jaundice, paralysis.

***Ethno- Medicinal uses :*** Seed - Stomach diseases (Shah & Joshi, 1971).

Stem - Jaundice (Joshi, 1982); eye diseases & wound (Jain & De, 1966); Indigestion & liver complaint (Sharma *et al.,* 1979); eczema (Deokule, 2006); diarrhoea, dysentery (Kataria & Gauchan, 2006).

Whole plant - Skin diseases, burns, wound (Goel *et al.,* 1984; Sharma *et al.,* 2006); fall of hair, diphtheria (Singh *et al.,* 1980); Bodyache (Kharkongor & Joseph, 1981; Sharma *et al.,* 2006); bone fracture (Krishna & Singh, 1987; Sharma *et al.,* 2006); cough (Kumar & Chauhan, 2005; Sharma *et al.,* 2006); fever (Rajwar, 1983; Sharma *et al.,* 2006); Itching (Goel & Mudgal, 1988; Sharma *et al.,* 2006); jaundice (Baruah *et al.,* 2006; Dutta Choudhury & Shil, 2006); swelling of the testicle (Prakash & Singh, 2006.); weakness, headache, indigestion, liver complaints, oedema (Sharma *et al.,* 2006).

## *CYMBOPOGON MARTINII* (Roxb.) Wats. (POACEAE)

***Syn.:*** *Andropogon martini* Roxb.

***Vern.:*** Rosha.

***Regional and Other Names :*** *English* - Rosha grass; *Hindi* - Gandhabel; *Kannada* – Harehullu; *Malayalam* –Sambharappullu; *Sanskrit* - Rohisah; *Tamil* - Kavattampillu; *Telugu* - Kamakchi–Kassuvu.

***Distribution :*** Throughout india.

***Taxonomic description :*** Tufed, sweet-scented perennial grass. Leaves amplexicaul at base. Panicles large, decompound; sessile spikeletes awned and lower glume have a deep median groove.

***Flowering & Fruiting :*** September– February.

***Medicinal parts :*** Whole plant.

***Taste :*** Bitter.

***Chemical constituents :*** Leaves contain Geranium oil.

***Medicinal properties :***

रोहिषं तुवरं तिक्तं कटुपाकं व्ययोहति ।
हृत्कण्डूव्याधि पित्तास्त्रशूलकासकफजवरान् ।। (भा.प्र.)

Rōhisaṃ tuvaraṃ tiktaṃ katupākaṃ vyapohati
hṛtkaṇdūvyādhi pittāṣraśūlakāsakaphajvarāṇ . (Bha. Pra.)

Acrid, appetiser, bitter, carminative, depurative, digestive, diuretic, febrifuge.

***Medicinal uses :*** Bronchitis, catarrh, cardiac debility, colic, cough, dyspepsia, fever, helminthiasis, leprosy, skin diseases.

***Ethno- Medicinal uses :*** Leaf - Rheumatism (Malhotra & Moorthy, 1973).

Oil - Skin disease & rheumatism (Sharma *et al.,* 1979).

## *CYNODON DACTYLON* (L.) Pers. (POACEAE)

***Syn.:*** *Panicum dactylon* L.

***Vern.:*** Dub.

***Regional and Other Names :*** *Bengali* – Durba; *English* - Doob grass, Burmunda grass; *Hindi* - Doob; *Kannada* – Kudigarikai; *Malayalam* – Arugampullu; *Marathi* – Harali; *Sanskrit*-Niladurva; *Tamil* - Arugampullu; *Telugu* – Gerichagaddi.

***Distribution :*** Cosmopolitan.

***Taxonomic description :*** Creeping, rhizomatous grass. Ligule a ring of hairs. Spikes 2-6, up to 6 cm long; lemmas longer than upper glumes, simple hairy. Grains turgid.

***Flowering & Fruiting :*** Most throughout the year.

***Medicinal parts :*** Whole plant.

***Taste :*** Little acrid & cooling.

***Chemical constituents :*** Plant contains β- Sitosterol, β- Carotene, Vitamin - C, Carotenes, Glucosides, Palmitic acid, Triterpenoid, Arundoin, Friedelene, Alkaloids – Erganovin, Erganovinine, ferulic, mono & oligo, Saccharide- glucose, fructose, sucrose, starch & lignin.

***Medicinal properties :***

नीलदूर्वा हिमा तिक्ता मधुरा तुवरा हरेत् ।
कफपित्तास्त्रवीसर्पतृष्णाद्राध्त्वगामयान् ।। (भा.प्र.)

Nīladurvā himā tiktā madhurā tuvarā hareṭ
Kaphapiṭtāṣravīsaṛpaṭrusṇādṛādhaṭvagāmayāṇ. (Bha. Pra.)

Astringent, cooling, constipating, depurative, diuretic, sweet.

***Medicinal uses :*** Abortion, Burning sensation, conjunctivitis, diarrhea, dysentery, general debility, haemorrhages, haemorrhoids, leprosy, skin diseases, vomiting, wounds.

***Ethno- Medicinal uses :*** Leaf - Dropsy (Bhalla *et al.,* 1982); wounds (Kumar *et al.,* 2006); piles (Bhalla et al., 1982; Kumar *et al.,* 2006).

Root - Dysentery, hysteria & piles (Nautiyal, 1981); chronic gleet (Upadhyay & Jaswant Singh, 2005).

Whole plant - Peptic ulcer (Harja & Baishya, 1981); bleeding piles (Deokule, 2006); diarrhoea (Kataria & Gauchan, 2006); dysentry (Kataria & Gauchan, 2006; Das *et al.,* 2006).

## *CYPERUS ROTUNDUS* L. (CYPERACEAE)

***Syn.:*** *C. hexastachys* Rottb.

***Vern.:*** Nagarmotha.

***Regional and Other Names :*** *Bengali* – Mutha; *English* - Nut grass; *Gujarati* - Motha; *Hindi* - Nagavmotha; *Kannada* – Tungegadde; *Malayalam* – Muttanna; *Marathi* – Nagarmotha; *Sanskrit* - Musta; *Tamil* - Korai; *Telugu* -Tunga musta.

***Distribution :*** Throughout India.

***Taxonomic description :*** Sedges, with slender stolons terminated by ellipsoid or globose-ovoid tubers. Leaves not exceeding the clum. Spikes born in terminal corymbs; spikelets dusty brown, 8 to 28 flowered; rachilla winged. Achenes triquetrous, brown.

***Flowering & Fruiting :*** Most of the year.

***Medicinal parts :*** Tubers.

***Taste :*** Acrid, bitter.

***Chemical constituents :*** Rhizome contains Essential oil which contains; Penene, cineol; fatty oil which contains; Linoleic acid, oleic acid, Myristic acid, Stearic; Glycerol, Glycoside cyperene, cyperenone, sugars, gum, carbohydrates.

Leaves contain Luteolin.

***Medicinal properties :***

मुस्तं हिमं कटु ग्राही तिक्तं दीपनपाचनम् ।
कषायं कफपित्तास्त्रतृड् ज्वरारूचि जन्तुहृत् ।। (भा.प्र.)

Mustaṃ himaṃ katu grāhī tiktaṃ dīpanapācanaṃ
Kasāyaṃ kaphapittāṣratṛḍ jvarāruchi jaṇtuhṛt. (Bha. Pra.)

Acrid, anthelmintic, anti – inflammatory, antiperiodic astringent, bitter, carminative, constipating, cooling, depurative, digestive, diuretic, expectorant, febrifuge, intellect promoting, stomachic.

***Medicinal uses :*** Amenorrhoea, amentia, bronchitis, colic, cough, diarrhoea, dysmenorrhoea, dysentery, dyspepsia, epilepsy, fever, flatulence, general calculi, general debility, inflammation, leprosy, pruritus, scabies, skin diseases, ulcer, verminosis, vomiting, wounds.

***Ethno- Medicinal uses :*** Bulb - Dysentery (Rao, 1981; Kataria & Gauchan, 2006); Jaundice (Rao, 1981); diarrhoea (Kataria & Gauchan, 2006); malaria (Deka *et al.,* 2006).

Root - Stomach pain (Maheshwari & Singh, 1984; Basak, 2006); snake bite (Mishra & Billore, 1983); jaundice (Gogol & Boissya, 1984).

## *DALBERGIA SISSOO* Roxb. (PAPILIONACEAE)

***Vern.:*** Shisham.

***Regional and Other Names :*** *Bengali* – Shisu; *English* - Indian red wood; *Gujarati* - Sissom; *Hindi* - Shisham; *Kannada* – Agaru; *Malayalam* – Shisavi; *Marathi* – Agaru; *Sanskrit* - Simsapa; *Tamil* - Sissu; *Telugu* – Nukku.

***Distribution :*** Throughout India.

***Taxonomic description :*** Deciduous tree, with 3-5, acuminate leaflets. Flowers yellowish, in axillary panicles. Stamens 9.

***Flowering & Fruiting :*** March – August.

***Medicinal parts :*** Root, Leaves, bark, heartwood.

***Taste :*** Bitter.

***Chemical constituents :*** Plant contains Dalbergin, dalbergenone, methyl dalbergin, nordalbergin, Isodalbergin, dalbergi-phenol, tecterigenin, Isocaviudin, glucoside, Isocaviunin, essential oil which contain: Bisboline & nerolidol.

***Medicinal properties :***

शिंशिपा कटुका तिक्ता कषाया शोषहारिणी ।
उष्णवीर्या हरेत् मेदः कुष्ठश्वित्रवमिकृमीन् ।।
वस्तिरूग्व्रणदाहास्त्रवलासान् गर्भपातिनी । (भा.प्र.)

Śiṃśipā katukā tiktā kasāyā śosahāriṇī
Uṣnavīryā hareṭ medaḥ kuṣthaśviṭravamikṛmīṇ
Vaṣtirugvṛaṇadāhāṣṛavalāsāṇ gaṛbhapātinī. (Bhā. Pra.)

Root - Astringent, constipating.

Leaf - Anthelmintic, Bitter, constipating, digestive, diuretic, styptic.

Bark/heartwood - Abortifacient, acrid, anthelmintic, anti – inflammatory, antipyretic, appetiser, astringent, bitter, depurative, thermogenic.

***Medicinal uses :*** Root - Diarrhoea, dysentery.

Leaf - Buring sensation, colic, diarrhoea, dysentery, dyspepsia, gonorrhoea, menorrhoea, vomiting.

Bark/heartwood - Amenorrhoea, dysmenorrhoea, burning sensation, bronchitis, dysentery, dyspepsia, fever, gout, hiccough, inflammation, leprosy, leucoderma, scabies, sciatica, skin diseases, syphilis, ulcers, vomiting.

***Ethno- Medicinal uses :*** Leaf - Gonorrohoea (Das *et al.,* 1983); dysentery (Singh & Maheshwari, 1983; Kataria & Gauchan, 2006); Skin diseases (Rajwar, 1983); Spermatorrhoea (Kumar & Chauhan, 2005); blood purifier (Shah, 2005); diarrhoea (Kataria & Gauchan, 2006).

Stem - Dysentery & headache (Maheshwari & Singh, 1984); Leprosy (Sharma, 1979).

Wood- Leprosy (Basak, 2006).

## *DATURA METAL* L. (SOLANACEAE)

***Syn.:*** *D. alba* Nees.; *D. fastuosa* L. var. *alba* C. B. Clarke.

***Vern.:*** Kala dhatura.

***Regional and Other Names :*** *English* - Thorn apple; *Hindi* - Kaladhatura; *Kannada* – Unmatta; *Malayalam* – Unmattu; *Sanskrit* - Dhustura; *Tamil* - Ummattai; *Telugu* – Ummatta.

***Distribution :*** Throughout India.

***Taxonomic description :*** A shrubby herb with dichotomous branches. Leaves ovate, sinuate, unequal sided at the base. Flowers purple or white, large, funnel shaped. Persistent calyx. Capsule globose, prickled.

***Flowering & Fruiting :*** July – March.

***Medicinal parts :*** Whole plant.

***Taste :*** Bitter.

***Chemical constituents :*** Seeds & flowers contain Daturadiol, Daturalone, fastusine, β- sitosterol, hyosine, hyoscyamine, atropine, fastudin, fastunine, fastusic acid, fastusidine, fastusinine, daluranlone, citrostadienol, obtusifoliol, allantoin, caproic, β- linolic, linoleic acid, oteic acid, datumetine, datumetelin, vitamin- C, chlorogenic acid, fixed oil.

***Medicinal properties :***

धस्तुरः कटुरूष्णश्च कान्तिकार री व्रणार्तिनुत् ।
त्वग्दोषखर्ज्जूक कंडूतिज्वरहरी भ्रमपदः ।। (रा. नि.)

Dhaṣturaḥ katuruaṇaśca kāṇtikāri vṛaṇārttinuṭ
Ṭvagdosa khaṛjju kaṇdūtijvarahari bhṛamapradaḥ. (Bhā. Pra.)

Root - Anodyne, antispasmodic, narcotic.

Seed - Antispasmodic, aphrodisiac, narcotic.

***Medicinal uses :*** Leaf - Asthma, dandruff, epilepsy, lumbago, mumps, neuralgia, otalgia, sciatica, swellings.

Seed - Asthma, otalgia, skin diseases.

***Ethno- Medicinal uses :*** Root - Leprosy (Jain & De, 1966); boils, ringworm (Purohit *et al.*, 1985); fever, skin diseases (Rajwar, 1983); asthma (Singh *et al.*, 1980).

Fruit - Fever, skin diseases (Sudhakar & Rolla, 1985); Boils (Maheshwari & Singh, 1987).

Leaf - Whooping cough, gonorrhoea, bronchitis (Bhalla *et al.*, 1982); boils (Maheswari & Singh, 1984); Wound (Lal & Yadav, 1983; Rajendran & Rengamani, 2006); ringworm (Purohit *et al.*, 1985); cough (Bedi, 1978); tetanus, arthritis, menorrhagia, headache, asthma, paralysis, body pain (Subhangi & Patil, 2006); asthma (Bhalla *et al.*, 1982; Rajendran & Rengamani, 2006).

Seed - Piles, skin diseases, tumour, urticaria, fistula (Bhalla *et al.*, 1982); leprosy, wound (Jain & De, 1966)); ringworm (Purohit *et al.*, 1985); cough (Singh & Pandey, 1980); fever (Shah, 2005); diabetes (Ghosh, 2006).

## *DATURA STRAMONIUM* L. (SOLANACEAE)

***Syn.:*** *D. tatula* L.

***Vern.:*** Safed dhatura.

***Regional and Other Names :*** *Bengali* - Safeid–dhutura; *English* - Thorn apple, devil's apple; *Hindi* -Safed dhatura; *Kannada* – Ummatta; *Malayalam* - Ummata, matulam; *Marathi* – Dhotra; *Sanskrit* - Dhattura; *Tamil* - Ummattai, vellumattai; *Telugu* - Ummetta, dutturamu.

***Distribution :*** Throughout India.

***Taxonomic description :*** An erect, annual herb; leaves sinuate, lanceolate, acute at apex, narrowed at base, membranous, sparsely pubescent; flowers white; capsule erect, oblong, densely covered with long prickles.

***Flowering & Fruiting :*** September – December.

***Medicinal parts :*** Leaves, flowers, fruits, seeds.

***Taste :*** Bitter.

***Chemical constituents :*** Plant contains Hysocyamine, hyoscine, atropine, scopalamine, tigloidin, tropine, linoleic acid, legnoceric, myristic, palmitic acid, oleic, stearic acid, β- carbolines, compesterol, β-sitosterol, stigmasterol, β- amyrin, daturadiol, gramisterol, chlorogenic acid, datugen, datugenine, daturalactone, stramonolide.

***Medicinal properties :***

धत्तूरो मदवर्णाग्निवातकृत् ज्वरकुष्ठनुत् ।
कषायो मधुरस्तिक्तो युकालिक्षाविनाशनः ।।
उष्णो गुरुर्व्रणश्लेष्मकण्डुकृमिविषापहः । (भा.प्र.)

Dhaṭṭūrō madavarṇāgnivātakṛt jvarakuṣthaṇuṭ
Kasāyo madhuraṣtikto yukāliksvinaśanāḥ
Uṣṇo guruṛvṛaṇaśleṣmakaṇdukṛmivisāpahaḥ. (Bhā. Pra.)

Seed - Acrid, alexiteric, anthelmintic, bitter, emetic, febrifuge, heating, sharp taste, tonic.

***Medicinal uses :*** Seed - Biliousness, itching, jaundice, leucoderma, piles, skin diseases, ulcer.

***Ethno-Medicinal uses :*** Leaf - Toothache (Shah, 1982); hydrocele (Chhetri, 2005); Asthma (Shah, 1982; Deokule, 2006); earache, eye infection (Baruah *et al.,* 2006).

Seeds - Rheumatism (Dar *et al.,* 1984); stomach pain (Shah, 1982); allergy (Das *et al.,* 2006).

## *DESMODIUM GANGETICUM* (L.) DC. (PAPILIONACEAE)

***Syn.:*** *D. collinum* Roxb.; *Hedysarum gangetium* L.

***Vern.:*** Shalparni.

***Regional and Other Names :*** *Bengali* – Shalpani; *Gujarati* - Salwan; *Hindi* - Salpan, Sarivan; *Kannada* – Nabiyalabune; *Malayalam* - Orila, Pullati; *Marathi* – Salparni; *Sanskrit* -Salaparni; *Tamil* - Orila, Pulladi; *Telugu* – Gitanaram.

***Description :*** Throughout India.

***Taxonomic description :*** Undershrub; leaflets ovate-oblong; flowers purple-pink, clustered in racemes; pods linear, flat, falcate.

***Flowering & Fruiting :*** August – November.

***Medicinal parts :*** Roots.

***Taste :*** Bitter.

***Chemical constituents :*** Plant contains Pterocarpanoids, Gangetin, gengetinin, Desmodin, alkaloids, Isoflavonoid+, Phytoalexins, Isoflavones, genestein.

***Medicinal properties :***

शालिपर्णी रसे तिक्ता गुरुष्णा वातदोषनुत् ।
विषमज्वरमेहार्शः शोफसन्तापनाशिनी ।। (रा. नि.)

Śāliparnī rase tiktā guruṣṇā vātadosanuṭ
Visamajvaramehaŕsaḥ śophasantāpanāśinī. (Bhā. Pra.)

Root - Anti-Inflammatory, aphrodisiac, bitter, carminative, constipating, diuretic, expectorant, febrifuge, sweet, thermogenic.

***Medicinal uses :*** Root - Asthma, debility, bronchitis, cough, dysentery, dyspepsia, fever, gout, haemorrhoids, inflammation.

***Ethno- Medicinal uses :*** Root - Snake bite (Goel *et al.,* 1984); asthma, cough (Sudhakar & Rolla, 1985); dysentery (Rao, 1981); fever & vomiting (Sharma, 1979); diarrhoea (Basak, 2006).

Whole plant - Bronchitis & fever (Karnick, 1981).

## *DICHROSTACHYS CINEREA* (L.) W. & A. (MIMOSACEAE)

***Syn.:*** *Mimosa cinerea* L.

***Vern.:*** Vartuli.

***Regional and Other Names :*** *Gujarati* - Marud; *Hindi* - Kheri; *Kannada* – Odatare; *Malayalam* – Vitattal; *Marathi* – Segumkati; *Sanskrit* - Vellantarah; *Tamil* - Vidattalai; *Telugu* –Veluturuchettu.

***Distribution :*** N.-W. India, Central India, Rajputana, Deccan.

***Taxonomic description :*** Polygamous shrubs, with branches ending into spines. Pinnae 4-14 pairs and leaflets 12-24 pairs. Spikes yellow in upper half and rose-pink or purple below. Pods twisted torulose.

***Flowering & Fruiting :*** August – October.

***Medicinal parts :*** Root.

***Taste :*** Bitter.

***Medicinal properties :***

वेल्लन्तरो रसे पाके तिक्तस्तृष्णाकफापहा ।
मूत्राघाताश्मजित् ग्राही योनिमूत्रानिलार्तिजित् ।। (कै. नि.)

Veḷlaṇtarō rase pāke tiktaṣṭrsṇākaphāpahā
Mūṭrāghātāsmajiṭ gṛāhi yonimūṭrānilāṛtijiṭ. (Kai. ni.)

Acrid, anodyne, anti-inflammatory, astringent, bitter, constipating, digestive, diuretic, thermogenic.

***Medicinal uses :*** Diarrhoea, dyspepsia, elephantiasis, inflammation.

***Ethno- Medicinal uses :*** Leaf - Leucorrhoea, menorrhoea (Hemadri & Rao, 1983).

## *DIOSCOREA ALATA* L. (DIOSCOREACEAE)

***Vern.:*** Chupri alu.

***Regional and Other Names :*** *Bengali* - Chupri alu; *English* - Great yam; *Hindi* - Chupri alu; *Kannada* – Onthalaigasu, Tengugenasu; *Malayalam* - Kachil, Kazhangu; *Sanskrit*-Arukam; *Tamil* - Perumvallikizhangu; *Telugu* – Pendalamu.

***Distribution :*** Only Known in Cultivation.

***Taxonomic description :*** Climbing or twining herbs, with large tuberous roots; stem 4- angled. Leaves cordate-ovate to cordate-oblong. Male flowers green, in panicles; female flowers in simple spikes. Capsules leathery. Elliptic. 3- Winged.

***Flowering & Fruiting :*** August – September.

***Taste :*** Bitter & Sweet.

Chemical constituents : ***Tuber contains Starch, Vitamin – B, Calcium oxalate, Protein, Iron.***

***Medicinal properties:***

आरूकं वीरसेनं च वीरं वीरारूकं तथा ।। (भा. प्र.)

Ārukaṃ vīrasenaṃ ca vīraṃ vīrārukaṃ tathā. (Bhā. Pra.)

Anthelmintic, aphrodisiac, cooling, diuretic, sweet, tonic.

***Medicinal uses :*** Diabetes, gonorrhoea, helminthiasis, leprosy.

***Ethno-Medicinal uses:*** Tuber-Abdominal pain (Tarafder, 1983); hydrocoel (Trebedi *et al.,* 1985); swelling (Saxena, & Vyas, 1983); piles (Das *et al.,* 2006).

## *DIOSCOREA BULBIFERA* L. (DIOSCOREACEAE)

***Syn.:*** *D. versicolor* Buch. Ham.; *D. pulchella* Roxb.; *D. sativa* Thunb.; *C. crispata* Roxb.

***Vern.:*** Varahi kand.

***Regional and Other Names :*** *Bengali* – Banalu; *English* - Patato yam; *Gujarati* - Suariya; *Hindi* - Ratalu; *Kannada* – Heggenasu; *Malayalam* - Kaltu kanchil; *Marathi* – Manakund; *Sanskrit* -Varahi; *Tamil* - Kodikilangu; *Telugu* – Chedupaddudumpa.

***Distribution :*** Throughout india.

***Taxonomic description :*** Slender, glabrous, twining herbs with globose to pyriform tubers; leaves broadly ovate, glabrous, acuminate, with a deeply cordate base; male flowers in raceme like, axillary spikes, greenish-purple; female flowers sessile, axillary or in fascicled spikes, purple; capsulc quadrately oblong.

***Flowering & Fruiting :*** August – October.

***Medicinal parts :*** Tubers, fruits.

***Taste :*** Bitter & Sweet.

***Chemical constituents :*** Plant contains A poisonous glucoside (Dioscorine).

Tuber contains Starch, Vitamin – B, Calcium oxalate, Protein, Iron.

***Medicinal properties :***

वाराही तु रसे स्वाद्वी तिक्ता पाके पुनः कटुः ।
शुक्रायुः स्वरवर्णाग्निबलपित्तविवर्द्धिनी ।
कफ कुष्ठमरुन्मेहकृमिहृन्च रसायनी ।। (भा.प्र.)

Vārāhī tu rase ṣvādvī tiktā pāke punaḥ katuḥ
Sukṛāyuḥ ṣvarvarṇāgnibalpiṭtavivaṛdhanī
Kaph kuṣtha marunmehkṛumiṛaṇca rasāyanī. (Bhā. Pra.)

Tuber - Alternative, anthelmintic, appetiser, aphrodisiac, bitter, fattening, pungent, stomachic, tonic.

***Medicinal uses :*** Tuber - Biliousness, bronchitis, dyspepsia, leucoderma, piles, strangury, tumors, urinary discharges.

***Ethno- Medicinal uses :*** Tuber - Abdomen pain, bone fracture (Joshi *et al.*, 1980); Boils (Jain & De, 1966); dysentery, piles (Sharma *et al.*, 1979); Jaundice (Krishna & Singh, 1987).

## *DIPLOCYCLOS PALMATUS* (L.) Jeffrey (CUCURBITACEAE)

***Syn.:*** *Bryonia laciniosa* L.; *B. palmata* L.; *Bryonopsis laciniosa* (L.) Naud.

***Vern.:*** Shivlingi.

***Regional and Other Names :*** *Bengali* – Mala; *Gujarati* - Shvalingi; *Hindi* - Shivlingi; *Kannada* – Lingatondi; *Malayalam* – Sivalingakkaya; *Marathi* – Shivalugi; *Sanskrit* - Lingini; *Tamil* -Sivalingakkaay; *Telugu* – Lingadauda.

***Distribution :*** Throughout India.

***Taxonomic description :*** Climbers, with 2-fid tendrils and 5-lobed leaves. Flowers greenish- yellow. Fruits globose, 6-striped; seeds with raised projections on both faces.

***Flowering & Fruiting :*** September- November.

***Medicinal parts :*** Whole plant.

***Chemical constituents:***

*Seed oil* Punicic acid.

***Medicinal properties :***

लिंगनी कटुरूष्णा च दुर्गन्धा च रसायनी ।
सर्वसिद्धिकरी दिव्या वश्या रसनियामिनि ।। (रा.नि.)

Liṇginī katuruṣṇā ca duṛgaṇdhā ca rasāyanī
Saṛvasiḍdhikari divyā vaśyā rasaniyāmini." (Ra.ni.)

Acrid, alterant, anti – Inflammatory, depurative, thermogenic.

***Medicinal uses :*** Cough, flatulences, general debility, inflammation, skin diseases.

***Ethno- Medicinal uses :*** Leaf - Eye infection (Goel *et al.,* 1984); body pain (Arinathan *et al.,* 2006).

Seeds - Promote fertility in women (Jain, 1963; Jadhav, 2006).

Root- Dysentery (Deokule, 2006).

## *DRYPETES ROXBURGHII* (Wall.) Hurus. (EUPHORBIACEAE)

***Syn.:*** *Putranjiva roxburghii* Wall.

***Vern.:*** Putranjiva.

***Regional and Other Names :*** *Bengali* - Putranjiva, Jiyapota; *English* - Wild olive; *Hindi* - Putranjiva; *Kannada* – Amani; *Malayalam* - Pongalam, Puttilanni; *Marathi* – Putajan; *Sanskrit* - Putrajivah; *Tamil* - Karupalai, Kattuppacceri; *Telugu* - Kundrajuvi, Kuduru.

***Distribution :*** Throughout India.

***Taxonomic description :*** Evergreen trees, with pendent branches and elliptic-oblong leaves. Flowers dioecious, apetalous, fascicled in axillary clusters. Fruits ellipsoid, hard.

***Flowering & Fruiting :*** March – May.

***Medicinal parts :*** Leaf, seed.

***Taste :*** Sweet, acrid.

***Chemical constituents :*** Fruit contains Manitol, Saponins.

Seed contains Arachidic acid, Linoleic acid, Palmitic acid, Stearic acid, Glucochlearin, Glucocleomin, Gluco Putranjivin, Putranjivoside, Putranoside, β- Sitosterol.

Leaves contain Stigmasterol.

Bark contains Carboxylic acid, Putric acid, Putranjivic acid.

***Medicinal properties :*** -

पुत्रजीवो गर्भकरो यष्टीपुष्पोऽर्थसाधकः ।। (भा.प्र.)

Putrajīvo gaṛbhakaro yaṣṭīpuṣpoaṛthasādhakaḥ. (Bhā. Pra.)

Leaf - Procreant, refrigerant.

Seeds - Acrid, anti – inflammatory, aphrodisiac, diuretic, procreant, laxative, sweet.

***Medicinal uses :*** Leaf - Catarrh, fever, sterility.

Seed - Azoospermia, burning sensation, constipation, elephantiasis, inflammation, sterility.

***Ethno-Medicinal uses:*** Fruit-Cold, fever, antifertility (Sharma *et al.*, 1979).

Seeds - Acidity & urtica (Hembrom & Goel, 2005).

## *ECHINOPS ECHINATUS* Roxb. (ASTERACEAE)

***Vern.:*** Ontkata.

***Regional and Other Names :*** *Bengali* – Thakurkanta; *English* - Thistle; *Gujarati* - Shuliyo; *Hindi* - Ontkatara; *Marathi* – Kadechubak; *Sanskrit* – Utkantaka, Kantalu.

***Distribution :*** Throughout India.

***Taxonomic description :*** Thistle-like annual herbs, with pinnatifid, spinescent leaves white woolly beneath. Heads white, blue or yellow, 1-flowered clustered in spinos globose compound heads.

***Flowering & Fruiting :*** October – March.

***Medicinal parts :*** Root.

***Taste :*** Bitter.

***Chemical constituents:***

Plant Hentriacontane, hentriacontanol, B-amyrin and leupeol.

Seeds Echinopsine.

***Medicinal properties :***

उत्कंटकः कटुस्तिक्तः कफवातहरी लघुः ।
तन्मूलपानतः स्त्रीणां शीघ्र प्रसवकारकः ।। (नि. र.)

Uṭkaṇtakaḥ katuṣtiktaḥ paphavātaharī laghuḥ
Taṇmulapānataḥ ṣṭrīnāṃ sighṛa pṛasavakārkaḥ. (Ni. ra.)

Abortifacient, aphrodisiac, bitter, cooling, pungent, sweet.

***Medicinal uses :*** Biliousness, diseases of the heart, gleet, strangury, urinary discharges.

***Ethno- Medicinal uses :*** Whole plant - Fever, malaria (Saxena & Vyas, 1981).

Leaf - Fever (Sharma *et al.,* 1979).

Root - Skin diseases (Maheshwari *et al.,* 1986); cold & sore throat (Jadhav, 2006).

## *ECLIPTA PROSTRATA* (L.) L. (ASTERACEAE)

***Syn.:*** *E. alba* Hassak.

***Vern.:*** Bhangra.

***Regional and Other Names :*** *Bengali* – Keshori; *English* - Trailing eclipta; *Gujarati* - Bhangra; *Hindi* - Bhangra; *Kannada* - Garagada soppu; *Malayalam* – Kannunni; *Marathi* – Bhringuraja; *Sanskrit* - Bhringraj; *Tamil* - Kayyantakara; *Telugu* – Galagara.

***Distribution :*** Bengal, Burma, Central India, Punjab, W. Rajputana, Peninsular India.

***Taxonomic description :*** Erect or prostate herbs, often rooting at the lower nodes. Leaves elliptic-oblong. Heads white, solitary.

***Flowering & Fruiting :*** Throughout the year.

***Medicinal parts :*** Whole plant.

***Taste :*** Bitter.

***Chemical constituents :*** Plant contains Glucosides – phytosterol – A, β-Amyrin, wedelolactone, luteoline, triterpenic acid, fatty acids, palmitic acid, stearic acid, oleic acid, linolic acid, nicotine, ampliptene alkaloid, ecliptine & resin.

***Medicinal properties :***

भृंगारः कटुकस्तिक्तो रूक्षोष्णः कफवातनुत् ।
केश्यस्त्वच्यः क्रमिश्वासकासशोथाम पाण्डुनुत् ।। (भा.प्र.)

Bhṛṅgārah katukastiktoruksoṣṇaḥ kaphavātanuṭ
Keśyaṣtvacyaḥ kṛmisvāsakāsasośthama paṇdunuṭ. (Bhā. Pra.)

Acrid, alterative, anodyne, anthelmintic, anti- inflammatory, aphrodisiac, bitter, carminative, depurative, digestive, diuretic, febrifuge, thermogenic.

***Medicinal uses :*** Debility, elephantiasis, fever, hypertension, inflammation, jaundice, leprosy, otalgia, pruritus, skin diseases, ulcers, wounds.

***Ethno- Medicinal uses :*** Root - Snake bite (Goel *et al.,* 1984); ulcer (Das *et al.,* 1983).

Whole plant - Elephantiasis, skin diseases, swelling, toothache (Bhalla *et al.,* 1982); ulcer, Jaundice (Rajwar, 1983); leucoderma (Shah *et al.,* 1983); Itching (Sexena & Dutta, 1975); asthma, bronchitis (Shah *et al.,* 1981); paralysis (Kumar & Chauhan, 2005); diarrhoea, dysentery (Kataria & Gauchan, 2006).

Leaf - Fever, malaria (Goel *et al.,* 1984); conjunctivitis (Banerjee, 1977); scorpion sting (Das *et al.,* 1983); gastric problem (kumar *et al.,* 1980); skin dieases (Purohit *et al.,* 1985); ulcer (Singh & Pandey, 1980); dysentery (Purkayastha & Subhan, 2006); blood purifier, diarrhoea, dysentery (Kiruba *et al.,* 2006); wound (Yesodharan & Sujana, 2006); diarrhoea (Purkayastha & Subhan, 2006; Arinathan *et al.,* 2006); elephantiasis (Sharma *et al.,* 2006).

## ***EMBELIA RIBES*** Burm. f. (MYRSINACEAE)

***Syn.:*** *E. glandulifera* Wt.

***Vern.:*** Baibirang.

***Regional and Other Names :*** *Bengali* – Baibirang; *English* - Barbreng; *Hindi* - Baibirang; *Kannada* - Vayavilanga; *Malayalam* – Vizhal; *Marathi* – Kakannie; Sanskrit- Vidanga; *Tamil* - Vayuvilanga; *Telugu* – Vayuvilanga.

***Distribution :*** Throughout India.

***Taxonomic description :*** A Large scandent shrub, bark rough, branches long thin with long internodes, Leaves cariaceous, elliptic lanceolate; flowers white or yellowish light green.

***Medicinal parts :*** Fruit.

***Taste :*** Acrid.

***Chemical constituents :*** Fruit contains Embelinn Quercitol compound, Volatile oil, Tannins, Alkaloids: Christembrine, Embelic acid, Resinoid.

***Medicinal properties :***

विडंग कटु तीक्ष्णोष्णं रूक्षं वहनिकरं लघु ।
शूलाध्मानोदर श्लेष्मकृमिवात विबन्धनुत् ।। (भा. प्र.)

Vidaṇgam katu tīkṣnosnam ruksyaṃ vahnikaraṃ laghu
Śulādhmānodara śleṣmakṛimivāta vibaṇdhanuṭ . (Bhā. Pra.)

Anthelmintic, Alterative, Blood purifier, rejuvinator, stomachic, tonic.

***Medicinal uses :*** Asthma, colic, cough, dyspepsia, epilepsy flatulance, general debility, indigestion, leprosy, skin diseases, toothache.

***Ethno- Medicinal uses :*** Fruits - Stomach complaint (Bhattacharjee *et al.,* 1980); toothache (Mokat & Deokule, 2006).

Seeds - Leprosy, ringworm, skin diseases (Purohit *et al.,* 1985).

Leaves- Gout (Jadeja *et al.,* 2006).

## *EMBELIA TSJERIAM-COTTAM* (R. & S.) DC. (MYRSINACEAE)

***Syn.:*** *E. robusta* C.B. Clarke

***Vern.:*** Bayabirang.

***Regional and Other Names :*** *English* - Basal; *Hindi* - Bayabirang; *Malayalam* - Cheriyannattam; *Marathi* – Ambati; *Sanskrit* – Birangbhaida.

***Description :*** From Ceylon and the Malabar coast through Sylhet and Assam to Singapore.

***Taxonomic description :*** Deciduous shrubs or small trees; leaves broadly elliptic; fruit red, globose.

***Flowering & Fruiting :*** February – June.

***Medicinal parts :*** Fruits.

***Taste :*** Bitter.

***Medicinal properties :***

विडंग कटु तीक्ष्णोष्णं रूक्षं वहिन्करं लघु ।
शूलाध्मानोदर श्लेष्मकृमिवात विबन्धनुत् ।। (भा. प्र.)

Vidaṇgam katu tīkṣnosnam ruksyaṃ vahnikaraṃ laghu
Śulādhmānodara śleṣmakṛimivāta vibaṇdhanuṭ . (Bhā. Pra.)

Fruits - Carminative, anthelmintic, antispasmodic.

***Medicinal uses :*** Fruits - Piles, tuberculous glands.

***Ethno- Medicinal uses :*** Leaf - Itching (Jain, 1963).

Bark - Burns (Joshi *et al.*, 1980); throat complaint, tonsil (Tribedi *et al.*, 1982).

Fruits - Blood purifier (Jain, 1963); fever (Tarafder, 1986).

## ***ENICOSTEMMA LITTORALE*** Blume. (GENTIANACEAE)

***Syn.:*** *E. axillare* (Lam.) Raynal.

***Vern.:*** Nau.

***Regional and Other Names :*** *Bengali* - Nagajivha; *Gujarati* - Mamejavi; *Hindi* - Chota chirayata; *Kannada* - Karibandita; *Malayalam* - Vellaruku; *Marathi* – Kadavinayi; *Sanskrit* - Nagajihva; *Tamil* - Vallari; *Telugu* - Chevvu – kurti.

***Distribution :*** Throughout India.

***Taxonomic description :*** Perennial herb, with 4-angular branches and linear-lanceolate leaves. Flowers white, in axillary clusters. Capsules 1-celled, 2-valved.

***Flowering & Fruiting :*** July – October.

***Medicinal parts :*** Whole plant.

***Taste :*** Bitter.

***Chemical constituents :*** Swertiamarin, gentianine, apigenin, genkwanin, Isovitexin, saponarin, heptacosane, nonacosane, myristic acid, oleic acid, gentiocrucine, alkaloids.

***Medicinal properties :***

नाहौ च नागजिहवाख्या तीक्ष्णपत्रा वितीक्ष्णिका ।
कृमिहृत् क्षारकर्मा च तथा मामज्जकः स्मृतः ।। (सो.नि.)

Nāhō ca nāgjivhākhyā tīkṣhaṇpaṭrā vitīkṣhaṇikā
Kṛumihṛt kshārkaṛmā ca tathā māmajjakaḥ ṣmṛuṭḥ. (So.ni.)

Acrid, anthelmintic, anti – inflammatory, antiperiodic, bitter, carminative, depurative, digestive, stomachic, thermogenic.

***Medicinal uses :*** Constipating, dyspepsia, dropsy, fever leprosy, pruritus, skin diseases, swelling.

***Ethno- Medicinal uses :*** Leaf - Fever (Joshi, 1982; Jadhav, 2006); worm infection (Pereira, 2006); malaria (Narain & Singh, 2006).

Whole plant - Fever & snake bite (Singh & Pandey, 1980); Malaria (Saxena, 1986); rheumatism, blood purifier (Kiruba *et al.,* 2006).

## *ERYTHRINA VARIEGATA* L. (PAPILIONACEAE)

***Syn.:*** *E. indica* Lam.

***Vern.:*** Paribhadra.

***Regional and Other Names :*** *Bengali* - Palita madar badisa; *English* - Indian coral tree; *Gujarati* – Bangaro; *Hindi* - Dadap, pankara; *Kannada* – Varjipe *Malayalam* – Murukku; *Marathi* - Pangara, harivana; *Sanskrit* – Paribhadrah; *Tamil* – Murukku; *Telugu* - Baridamu, bodita.

***Distribution :*** Coast forests from Bombay to Malabar and in the Andamans and Nicobars; much planted for ornament.

***Taxonomic description :*** Small tree, with 3-foliate leaves, red flowers and black, torulose pods.

***Flowering & Fruiting :*** February – May.

***Medicinal parts :*** Bark, Leaves.

***Taste :*** Acrid.

***Chemical constituents :*** Bark contains A bitter erytherine alkaloid & resins.

Leaves contain Erytherine, alkaloid.

***Medicinal properties :***

परिभद्रः कटुष्णः स्यात् कफवातनिकृन्तनः ।
अरोचकहरः पटयो दीपनश्यापि कीर्तितः ।। (रा. नि.)

Paribhadraḥ katuṣṇaḥ ṣyāṭ kaphavātanikṛṇtanaḥ
Arocakaharaḥ pathyo dīpanaśyāpi kīṛtitaḥ. (Raj. ni.)

Bark - Acrid, anthelmintic, anti – inflammatory, bitter, carminative, depurative, digestive, diuretic, expectorant, febrifuge, rejuvenating, sedative, stomachic, thermogenic.

Leaf - Anthelmintic, bitter, diuretic, laxative, stomachic.

***Medicinal uses :*** Bark - Colic, conjunctivitis, cough, dyspepsia, fevers, inflammation, insomnia, leprosy, otalgia, skin diseases.

Leaf - Dysmenorrhoea, otalgia.

***Ethno-Medicinal uses :*** Leaf - Cough (Jain & De, 1966); ulcer (Apparanantham & Chelladurai, 1986); toothache (Aditya *et al.,* 2006).

Bark - Cold, cough, eye complaint (Jain & De, 1966); fever, rheumatism, skin diseases (Nautiyal, 1981).

## *EUPHORBIA ANTIQUORUM* L. (EUPHORBIACEAE)

***Vern.:*** Tidhara.

***Regional and Other Names :*** *Bengali* - Tiktasij, Tesiramonsa, bajwaran; *English* - Triangular spruge; *Gujarati* -Tandharisend; *Hindi* - Tidhara; *Kannada* - Mudumula, Mulajemudu; *Malayalam* -Caturakkalli, Narasya; *Malayalam* – Chaudhariniwdung; *Sanskrit* - Vajrakantakah; *Tamil* - Saturakkalli, Vachirom; *Telugu* - Bommajemudu, Bonthakali.

***Distribution :*** Throughout India.

***Taxonomic description :*** Much branched shrubs or a small trees; branches curving upwards, stout, fleshy, jointed, 3-angled; leaves subsessile, obovate-oblong or spathulate, fleshy, long, Deciduous; involucres 3-nate, forming small pedunculate cymes, central flowers sessile, lateral stalked; cocci compressed, glabrous.

***Flowering & Fruiting :*** July – September.

***Medicinal parts :*** Root, Juice.

***Taste :*** Bitter.

***Chemical constituents :*** Plant contains Euphol, Taraxerol, Epifriedelanol, friedelan.

Latex cntains β-Amyrin, antiquorin, cycloartenol, œ-euphorbol, euphadienol.

***Medicinal properties :***

स्नुही चान्या त्रिधारा स्यात् तिस्त्रो धारास्तु यत्र सा ।
पूर्वोक्तगुणवत्येषा विशेषाद् रससिद्धिदा ।। (रा. नि.)

ṣnuhi Cāṇyā ṭridhārā ṣyāṭ tiṣṭro dhārāṣtu yaṭra sā
Puṛvoktagunavaṭyesā viśesāḍ rasasiḍdhidā. (Ra. Ni.)

Root - Acrid, anodyne, bitter, digestive, emetic, purgative, stomachic, thermogenic.

Juice - Acrid, anti-Inflammatory, purgative.

***Medicinal uses :*** Root - Colic, constipation, dyspepsia, ulcers, wounds.

Juice - Cough, deafness, dropsy, gout, rheumatism.

***Ethno- Medicinal uses :*** Latex/Juice - Burns (Borthakur, 1976); rheumatism (Rai Chaudhuri & Pal, 1976); Dropsy (Deokule, 2006).

Whole plant - Skin diseases (Mishra & Sahu, 1984).

Root- Toothache (Aditya *et al.,* 2006).

## *EUPHORBIA HIRTA* L. (EUPHORBIACEAE)

***Syn.:*** *E. pilulifera* L.

***Vern.:*** Dudhi.

***Regional and Other Names :*** *Bengali* - Baro kheruie; *English* - Australian asthma herb; *Gujarati* - Dudeli; *Hindi* - Dudhi; *Malayalam* - Nela palai; *Marathi* – Dudhi;*Sanskrit* -Dugdhika; *Tamil* - Pallavi; *Telugu* – Pacchaku.

***Distribution :*** Throughout India.

***Taxonomic description :*** Erect or procumbent herbs, with ovate-lanceolate or rhomboid-oblong leaves. Cyathia greenish, crowded in pedunculate cymes; glands stalked, pink. Capsules globosely 3-gonous.

***Flowering & Fruiting :*** Throghout the year.

***Medicinal parts :*** Whole plant.

***Taste :*** Bitter.

***Chemical constituents :*** Plant contains Quercetin, Triacentane, ellagic acid, Phenolics, Euphosterol, Phytosterol, Phytosterolin, Gallic, mellistic, palmitic, oleic, linoleic acid, zanthorhamnin wax, chlorophyll, resin, tannins, sugar, mucilage, carbohydrates, albuminoids, catchin, oxalate.

***Medicinal properties :***

दुग्धिकोष्णा गुरु रूक्षा वातला गर्भकारिणी । (भा.प्र.)

Dugdhikoṣnā guru rukshā vāttalā gaṛbhkāriṇī. (Bhā. Pra.)

Astringent, diuretic, expectorant.

***Medicinal uses :*** Asthma, colic, cough, dysentery, gonorrhoea, intestinal worms in children, leucoderma.

***Ethno- Medicinal uses :*** Leaf - Ringworm, scabies (Shah *et al.*, 1981); burn, pain in joints (Jain *et al.,* 1973); gonorrhoea (Mishra & Sahu, 1984); snake bite (Goel *et al.*, 1984); dysentery (Ganeshan *et al.,* 2006); wound (Yesodharan & Sujana, 2006).

Latex - Eczema, scabies (Bhatt & Sabnis, 1987); eye complaint (Bhalla *et al.,* 1982); ringworm (Joshi, 1982); scorpion bite (Maheshwari *et al.,* 1986); wort (Mishra & Sahu, 1984); pimples (Kiruba *et al.,* 2006); skin irritation (Gupta, 2006).

Whole plant - Colic (Maheshwari & Singh, 1984); dysentery, cough (Mishra & Sahu, 1984; Kataria & Gauchan, 2006); Cuts (Saxena & Dutta, 1975); diarrhoea (Audichya *et al.,* 1983; Kataria & Gauchan, 2006); toothache (Sebastian & Bhandari, 1984); vomiting (Sexena & Vyas, 1983); kidney diseases (Sudhakar & Rolla, 1985); leucorrhoea (Kumar & Chauhan, 2005); dysentery (Upadhyay & Jaswant Singh, 2005); lucorrhoea & spermatorrhoea (Kumar & Chauhan, 2005); cough (Deokule, 2006); chronic bronchial affections, asthma (Mishra & Sahu, 1984; Kataria & Gauchan, 2006; Kumar *et al.,* 2006); stomach trouble in children (Rajendran & Rengamani, 2006).

## *EUPHORBIA NERIIFOLIA* L. (EUPHORBIACEAE)

***Vern.:*** Ghota thuwar.

***Regional and Other Names :*** *Bengali* – Monshasij; *English* - Holy milk hedge; *Gujarati*-Thor; *Hindi*-Sij, shend; *Kannada*–Yalekalli; *Malayalam*–Illakka- lli; *Marathi*–Nevagunda; *Sanskrit*-Sehund; *Tamil*-Illakkalli; *Telugu*–Akejemudu.

***Distribution :*** W. Peninsula Cultivated elsewhere and in Ceylon.

***Taxonomic description :*** Erect deciduous shrubs or small trees; stem 5-angled. Leaves obovate-oblong, crowded at the end of stem, subfleshy.

Involucre 3-nate; glands without petaloid limb. Capsules obovoid-trigonous, deep red or pale brown.

***Flowering & Fruiting :*** February– March.

***Medicinal parts :*** Roots, stem, leaves.

***Taste :*** Bitter.

***Chemical constituents :*** Plant contains Euphol, taraxerol, friedelan, resin, gum.

Bark contains n- Hexacosonal, Euphorbol, Hexacos

Anoate, Pelargonine, diglusoside, Euphorbin.

Root & bark contains Methylene Cycloartenol, Tulipanin 3 -5 diglucoside.

***Medicinal properties :***

सहूण्डो रेचनस्तीक्ष्णो दीपनः कटुको गुरुः ।
शुलामाष्ठीलिकाऽऽध्मानकफगुल्मोदरानिलान् ।।
उन्मादमोहकुष्ठार्शः शोथमेदोऽश्मपाण्डुताः ।
व्रणशोथज्वरप्लीहविषदूषीविषं हरेत् ।। (भा.प्र.)

Sehūṇdo racanaṣtīkṣṇo dīpanaḥ katuko guruḥ
Śulāmāṣthīlikāadhmānkaphaguḷmodarānilāṇ
Uṇmādmohkuṣthārśhaḥ śothmadoaśmpāṇdutahāḥ
Vṛanśothjvarpḷīhavisdūsīvisaṃ hareṭ. (Bhā. Pra.)

Plant - Alexipharmic, bitter, carminative, laxative, pungent.

Milk - Appetiser, laxative, pungent.

Leaf - Appetiser, carminative.

***Medicinal uses :*** Plant - Abdominal trouble, anaemia, bronchitis, fever, inflammation, leucoderma, piles, tumours, ulcer.

Milk - Abdominal pain, leucoderma, tumor.

Leaf - Abdominal trouble, inflammation, swelling, tumour.

***Ethno- Medicinal uses :*** Milk - Sore throat (Jadhav, 2006); earache (Deokule, 2006).

leaf – Cough (Sarma *et al.,* 2006).

## *EUPHORBIA TIRUCALLI* L. (EUPHORBIACEAE)

***Vern.:*** Sehund.

***Regional and Other Names :*** *Bengali* - Lanka sij; *English* - Milk bush; *Gujarati* - Dandaliothora; *Hindi* - Sehund, Konpal; *Kannada* – Bontakalli; *Malayalam* – Kolkalli; *Marathi* – Vajraduhu; *Sanskrit* - Bahukshira; *Tamil* - Tirukkalli; *Telugu* - Chemundu, Pullakada.

***Distribution :*** Throughout India.

***Taxonomic description :*** Shrubs or small trees, with smooth, spreading, terete, rush-like, almost leafless branches.

***Flowering & Fruiting :*** August– September.

***Medicinal parts :*** Root, stem, latex, bark.

***Taste :*** Bitter.

***Chemical constituents :*** Stem & latex contain β- Sterol, Taraxerol.

Bark & root contain Euphorbol, Hexacosanoate, Taraxerone, Ingenol, Triacetate, Euphorginol.

Stem contain Ellagic acid, Kaempferol, compesterol, Stigma – Sterol.

Latex contain Amyrin, Euphol, Euphorbinol, Tirucallol, Palmitic acid.

***Medicinal properties :***

उष्णवीर्यं स्नुहीक्षीरं स्निग्धंच कटुकं लघु ।
गुल्मिनां कुष्ठिनाच्चापि तथैवोदररोगिणाम् ।
हितमेतद्विरेकार्थे ये चान्ये दीर्घरोगिणः ।। (भा. प्र.)

Uṣṇavīryaṃ ṣnuhīkshīraṃ ṣnigdhaṇch katukaṃ laghu
Guḷmināṃ kuṣthināccāpi tathevoderrōgiṇāṃ
Hitmetḍvirekāṛth ye chānye dīrgharōgiṇaḥ . (Bhā. Pra.)

Milk - Alexiteric, carminative, emetic, purgative, rubefacient.

***Medicinal uses :*** Plant - Biliousness, leprosy, leucorrhoea.

Milk - Abdominal trouble, tumour.

***Ethno-Medicinal uses :*** Latex-Rheumatism (Nagendra Prasad & Abraham, 1984; Deokule, 2006); toothache, warts, earache (Mishra & Sahu, 1984); eczema & scabies (Ramachandran & Nair, 1981); wounds (Shah *et al.*, 1983).

Stem- Wound (Pereira, 2006).

Root - Colic (Shah, 2005).

## *EVOLVULUS ALSINOIDES* L. (CONVOLVULACEAE)

***Syn.:*** *E. hirsutus* Lamk.

***Vern.:*** Nil Shankhpuspi.

***Regional and Other Names :*** *Gujarati* - Jhinkiphudari; *Hindi* - Shyamakranta; *Kannada* – Vishnukranti; *Malayalam* – Vishnukanthi; *Marathi* – Vishnukrant; *Sanskrit* - Laghuvishnukranta; *Tamil* - Vishnukirandi; *Telugu* – vishnkanta.

***Distribution :*** Throughout India.

***Taxonomic description :*** Prostrate or ascending herbs, with lanceolate-oblong leaves. Flowers pale-blue, on 1 to few flowered filiform peduncles. Calyx-lobes acute. Capsules 4-valved, 4-seeded.

***Flowering & Fruiting :*** July – March.

***Medicinal parts :*** Whole plant.

***Taste :*** Bitter.

***Chemical consltiluents :*** Plant contains Yellow fatty elements, Minerals, Organic acid & Saline.

***Medicinal properties :***

शंखपुष्पी सरा मेध्या वृष्या मानसरोगहृत् ।
रसायनी कषायोष्णा स्मृतिकान्ति बलाग्निदा ।।
दोषापस्मारभुताश्रीकुष्ठक्रिमिविषप्रणुत् । (भा.प्र.)

Śaṇkhapuṣpī sarā meḍhyā vṛṣyā mānasarogahṛt
Rasāyanī kasāyoṣṇā ṣmṛiti kāṇti balāgnidā
Dosāpṣmārbhutā sṛī kuṣtha ḳrimi visp.ṛaṇuṭ. (Bhā. Pra.)

Plant - Alexiteric, alterative, anthelmintic, bitter, pungent, tonic.

***Medicinal uses :*** Plant - Biliousness, bronchitis, epilepsy, leucoderma, teething in infants.

***Ethno- Medicinal uses :*** Leaf - Asthma, bronchitis (Sudhakar & Rolla, 1985); ulcer (Bhatt & Sabins, 1987); fever, dysentery, rheumatism (Kiruba *et al.,* 2006); improve memory (Rajendran & Rengamani, 2006).

Whole plant - Scorpion sting (Jain & De, 1966); asthma (Singh & Pandey, 1980); stomach pain, leucorrhoea (Maheshwari & Singh, 1987); dysentery (Sharma *et al.*, 1979; Kataria & Gauchan, 2006); fever (Deokule, 2006); diarrhoea (Kataria & Gauchan, 2006).

Root - Fever (Lal & Yadav, 1983).

## *FICUS BENGHALENSIS* L. (MORACEAE)

***Vern.:*** Bargad.

***Regional and Other Names :*** *Bengali* – Bot; *English* - Banyan tree; *Gujarati* - Vad; *Hindi* - Bargad, bar; *Kannada* - Mara, ala; *Malayalam* - Ala, vatam; *Marathi* - Vada, wad; *Sanskrit* -Vata; *Tamil* - Alam, vadam, al; *Telugu* - Vati, peddamarri.

***Distribution :*** Sub-Himalayan tract and W. Peninsula. Planted elsewhere.

***Taxonomic description :*** Large, evergreen trees, with prop roots and ovate-oblong leaves. Receptacles geminate, axillary, globose, red when ripe.

***Flowering & Fruiting :*** August – July.

***Medicinal parts :*** Whole plant, buds, roots, latex.

***Taste :*** Astringent.

***Chemical constituents :*** Plant contains Glucoside; Bengalenoside & Talbutamide; Leucoanthocyanidins, Phytosterolin, ß- sitosterol, freidelin, quercetin, Galactoside, ticlic acid, tannin.

***Medicinal properties :***

वटः शीतो गुरुग्राही कफपित्तव्रणापहः ।
वर्ण्यो विसर्पदाहघ्नः कषायो योनिदोषहृत् ।। (भा.प्र.)

Vataḥ śito gururgrāhī kaphpiṭtavṛāṇapaḥ
Vaṛṇyo visaṛpadāhghanḥ kasāyo yonidoshṛaṭ. (Bha.Pra.)

Acrid, astringent, sweet.

***Medicinal uses :*** Biliousness, erysipelas, fever, inflammation, ulcers, vaginal complaint, vomiting.

***Ethno- Medicinal uses :*** Bark - Snake bite (Singh & Pandey, 1980); diabetes (Negi *et al.,* 1985; Hembrom & Goel, 2005; Ghosh, 2006).

Aerial root - Diarrhoea (Jain & De, 1966); menorrhoea (Joshi *et al.,* 1980); leucorrhoea (Hemadri & Rao, 1983).

Leaf - Blisters, boils (Tarafder, 1984); cough (Singh & Pandey, 1980); piles (Kumar & Chauhan, 2005); cuts (Maheshwari & Singh, 1984).

Latex - Gum trouble & toothache(Nagendra Prasad & Abraham, 1984; Aditya *et al.,* 2006); sex diseases (Rai, 1985); eye complaint (Singh *et al.,* 1987); blisters, boils (Singh & Maheshwari, 1983); skin diseases (Maheshwari & Singh, 1987); rheumatic pain (Deokule, 2006); fever (Rajendran & Rengamani, 2006).

## *FICUS GLOMERATA* Roxb. (MORACEAE)

***Syn.:*** *F. racemosa* L.

***Vern.:*** Gular.

***Regional and Other Names :*** *Bengali* – Dumur; *Gujarati* – Umar; *Hindi* – Gular; *Malayalam* - Atti; *Marathi* – Gular; *Sanskrit* -Udumbara; *Tamil* - Adavam; *Telugu* – Atti.

***Distribution :*** Throughout India.

***Taxonomic description :*** A large tree; leaves alternate, ovate, elliptic-lanceolate, or ovate-oblong, glabrous. Receptacles 2-4 cm across, clustered on short leafless branchlets, globose or pyriform, hairy, supported by 3 bracts at the base, green, turning reddish-purple when ripe.

***Flowering & Fruiting :*** February – June.

***Medicinal parts :*** Whole plant, roots, leaves, bark, fruits, latex.

***Taste :*** Acrid, sweet.

***Chemical constituents :*** Bark contains Cerylbehenate, Lupeol, Alphaamyrin, Glaunol, β- sitosterol, keton, Hentriacontane, tiglic acid, glucose, Taraxeroll & Tannin.

***Medicinal properties :***

उदुम्बरो हिमो रूक्षो गुरुः पित्तकफास्त्रजित् । (भा.प्र.)

Udumbarō himo rukso guruḥ piṭtkaphaṣṭrijiṭ. (Bhā. Pra.)

Plant - Acrid, antidysenteric, cooling, sweet.

***Medicinal uses :*** Root - Hydrophobia.

Fruits - Biliousness, blood diseases, burning sensation, fatigue, leprosy, menorrhagia, nose bleeding, urinary discharges.

***Ethno- Medicinal uses :*** Latex - Dysentery (Jadhav, 2006); diarrhoea (Kumar & Chauhan, 2005); piles, diarrhoea (Deokule, 2006).

Bark- Piles (Pereira, 2006); dysentery (Deokule, 2006); diabetes (Kumar *et al.*, 2006).

Root- diarrhoea, dysentery (Kataria & Gauchan, 2006); diabetes (Kumar *et al.*, 2006).

Fruit – Diabetes (Kumar *et al.*, 2006).

## *FICUS RELIGIOSA* L. (MORACEAE)

***Vern.:*** Pipal.

***Regional and Other Names :*** *Bengali* – Ashathwa; *English* - Pipal tree; *Gujarati* - Pipro; *Hindi* - Pipal; *Kannada* - Ashwattha, arali; *Malayalam* – Ashvatham; *Marathi* – Ashvatha; *Sanskrit* -Pippala; *Tamil* - Aswattham; *Telugu* – Aswatthamu.

***Distribution :*** Sub-Himalayan forests, Bengal, Central India.

***Taxonomic description :*** Large, deciduous trees, with ovate to ovate-suborbicular leaves caudate-acuminate at apex and rounded at the base. Receptacles in axillary pairs, reddish-purple when ripe.

***Flowering & Fruiting :*** April – June.

***Medicinal parts :*** Bark, shoot, leaves, fruit.

***Taste :*** Sweet – astringent.

***Chemical constituents :*** Bark contains Tannin.

***Medicinal properties :***

पिप्पलो गुर्जरः शीतः पित्तश्लेष्मवृणास्त्रजित् ।। (भा.प्र.)

Pipplo guṛjarḥ śitaḥ piṭtaśleshamvṛaṇāṣtiṛjit. (Bhā. Pra.)

Plant - Acrid, bitter, sweet.

Fruits - Alexipharmic, cooling.

***Medicinal uses :*** Burning sensation, biliousness, diseases of the blood, uterus, vagina; leucorrhoea, ulcers.

Fruits - Biliousness, burning sensation, diseases of the blood & heart.

Root - gout

Root bark - Leucorrhoea, stomatitis, ulcers.

Seed - Urinary discharges.

***Ethno- Medicinal uses :*** Fruits - Asthama (Bhalla *et al.,* 1982).

Bark - Ulcer, boils, blisters (Maheshwari & Singh, 1984); scabies (Bhalla *et al.,* 1982); snake bite (Murthy *et al.,* 1986); whooping cough (Virjee *et al.,* 1984); gonorrhoea (Bhalla *et al.,* 1982; Purkyastha & Subhan, 2006; Basak, 2006).

Leaf - Ulcer, boils, blisters (Shah, 1982); toothache (Shekhawt & Anand, 1984); skin diseases (Malhotra & Moorthy, 1973); Snake bite (Maheshwari *et al.,* 1986); scabies (Deokule, 2006).

## *FUMARIA INDICA* (Hassk.) Pugsley (FUMARIACEAE)

***Syn.:*** *F. vaillantii* Loisel.; *F. parviflora* Lam.

***Vern.:*** Pitpapra.

***Regional and Other Names :*** *Bengali* – Bansalpha; *Hindi* - Pitpapra; *Marathi* – Pitpada; *Sanskrit* - Parpat; *Tamil* - Thura; *Telugu* – Chatharasi.

***Distribution :*** Over the Greater part of India.

***Taxonomic description :*** Annual herbs; leaves 2-3 pinnatisect, segments linear-oblanceolate; flowers purplish-red, spurred, in terminal or leaf opposed, bracteate racemes; sepals inciso-dentate; nuts subglobose, rugose when dry.

***Flowering & Fruiting :*** January – April.

***Medicinal parts :*** Whole plant.

***Taste :*** Bitter.

***Chemical constituents :*** Plant contains Fumaric acid & fumarin alkaloid.

***Medicinal properties :***

पर्पटोहन्ति पित्तास्त्रभ्रमतृष्णाकफज्वरान् ।
संग्राही शीतलस्तिक्तो दाहनुद्वातलो लघुः ।। (भा. प्र.)

Paṛpatohaṇti piṭtāṣṭrabhṛimṭriṣhṇākaphajwarāṇ
Saṇgrāhī śītalṣtikto dāhnuḍvātalo laghuḥ. (Bhā. Pra.)

Aperient, astringent, bitter, blood purifier, diaphoretic, diuretic, tonic.

***Medicinal uses :*** Blood diseases, fever, leprosy, rhinitis, tuberculousis, vertigo.

***Ethno- Medicinal uses :*** Seed - Bodyache (Sharma *et al.,* 1979).

Whole plant - Liver complaint, blood purifier (Sharma *et al.,* 1979); skin diseases (Lal & Yadav, 1983); flue (Shah & Joshi, 1971); fever (Maheshwari & Singh, 1987); bodyache (Singh, 1945; Hembrom & Goel, 2005); blood purifier (Shah, 2005); dysentery (Jadeja *et al.,* 2006).

## *GLORIOSA SUPERBA* L. (LILIACEAE)

***Syn.:*** *Methonia superba* Lamk.

***Vern.:*** Kalihari.

***Regional and Other Names :*** *Bengali* – Ulatchanadal; *English* - Malabar glory lily; *Gujarati* - Dudhi ovachhnag; *Hindi* - Karihari, langali; *Kannada* – Agnisikha; *Malayalam* - Mallattamara, mettonni; *Marathi* - Kariannag, indai; *Sanskrit* -Kalihari; *Tamil* - Kalaippaikokishangu; *Telugu* - Adavi nabhi, deyyapaku, kalappagadda.

***Distribution :*** Throughout India.

***Taxonomic description :*** Climbing herbs, with alternate, opposite or whorled leaves ending in a tendril. Flowers yellow, tinged with red in upper half; tepals undulating on the margins, Capsules oblong.

***Flowering & Fruiting :*** August– October.

***Medicinal parts :*** Rhizome, root.

***Taste :*** Bitter, acrid, pungent, astringent.

***Chemical constituents :*** Rhizome tubers contain Alkaloid colchicine; superbine, glorisine; sitosterol, glucosides, benzoic acid & resins.

***Medicinal properties :***

कलिहारी सरा कुष्ठशोफार्शोव्रणशुलजित् ।
सक्षारा श्लेष्मजितिक्ता कटुका तुवराऽपि च ।।
तीक्ष्णोष्णा कृमिहृल्लध्वी पित्तला गर्भपातिनी । (भा.प्र.)

Kalihārī sarā kuṣthaśophārśovṛanśuljịt
Saksārā śloṣmjitiktā katukā tuwrāapi ca
Trīkṣṇoṣṇā kṛmihṛllaghvī pịttalā gaṛbhapātinī. (Bhā. Pra.)

Tuber - Abortifacient, acrid, alexiteric anthelmintic, bitter, heating, laxative, Pungent.

***Medicinal uses :*** Tuber - Abdominal pain, infalammation, itching, leprosy, piles, ulcer.

***Ethno- Medicinal uses :*** Root - Anthelmintic (Sharma *et.al.,* 1979 )

Leaf - Asthma (Tarafder and Rai Chaudhuri, 1981); Gout (Jadeja *et al.,* 2006).

Rhizome - Scorpion bite, Skin diseases & Gonorrhoea (Singh *et.al.,* 1980); gout (Jain & Dam, 1979); Intermittent fever (Tarafder & Rai Choudhary, 1981); Abortifacient (Tarafder, 1983; Rajendran & Rengamani, 2006); Leprosy, piles (Singh & Pandey, 1980); Snake bite (Bedi, 1978); Stomach diseases (Baruah *et al.,* 1987); rheumatism (Chatterjee & Bhatterjee, 2006).

## *GLYCYRRHIZA GLABRA* L. (PAPILIONACEAE)

***Syn.:*** *G. gladulifera* W. & K.; *G. violacea* Boiss.

***Vern.:*** Mulethi.

***Regional and Other Names :*** *Bengali* – Jastimodhu; *English* - Liquorice; *Gujarati* - Jethimadha; *Hindi* - Mulethi; *Kannada* - Yashti madhuka; *Malayalam* - Iratimadhuram; *Marathi* - Jashtamadhu; *Sanskrit*- Yasti, Madhuka; *Tamil* - Atimaduram; *Telugu* - Atimadhuramu.

***Distribution :*** British Baluchistan. Europe, Mediterranean, N. Africa.

***Taxonomic description :*** A perennial small spreading shrub with pinnate compound leaves, purple flowers.

***Medicinal parts :*** Root, underground stem, leaves.

***Taste :*** Sweet.

***Chemical constituents :*** Plant contains Glycyrrhizin, glyciyrrhizic acid, It is 50 times sweeter than the normal sugar. Glycosides, steroids, glucose, sucrose, resin, starch and essential oil.

***Medicinal properties :***

यष्टी हिमा गुरुः स्वाद्वी चक्षुष्या बलवर्णकृत् ।
सुस्निग्धा शुक्रला केश्या स्वर्या पित्तानिलास्तजित् ।। (भा.प्र.)

Yastī himā guruḥ ṣvaḍvī caksuṣyā balavaṛnakṛt
Suṣnigdhā śukṛalā keśyā ṣvaṛyā piṭtānilāṣrajiṭ. (Bhā. Pra.)

Anti-emetic, anti-inflammatory, aphrodisiac, appitizer, blood purifying, cooling, deumulcent, diuretic, expectorant, haemostatic, rejuvenator, sweet, tonic, wound healing.

***Medicinal uses :*** Cholera, cough, epilepsy, haematemesis, heart diseases, retention of urine, skin diseases, spermopoitic, vomiting, wound healing.

***Ethno- Medicinal uses :*** Root - Throat paịn (Ganesan *et al.,* 2006).

## *GMELINA ARBOREA* Roxb. (VERBENACEAE)

***Syn.:*** *G. rheedii* Hook.

***Vern.:*** Gambhari.

***Regional and Other Names :*** *Bengali* – Gamari; *English* - White teak; *Gujarati* - Savan; *Hindi* - Gamhar; *Kannada* – Shivani; *Malayalam* – Kumbil; *Marathi* – Shewan; *Sanskrit* -Kasamari; *Tamil* - Kumadi; *Telugu* - Gummadi, Errinaruva.

***Distribution :*** Throughout India.

***Taxonomic description :*** Unarmed deciduous trees, with broadly ovate-acuminate leaves. Flowers yellow or brownish-yellow. Drupes obovoid, yellowish orange when ripe.

***Flowering & Fruiting :*** February – April.

***Medicinal parts :*** Whole plant, roots, leaves, fruits.

***Taste :*** Bitter, acrid.

***Chemical constituents :*** Plant contains Lignan gmelinol & Arboreol.

Leaves contains Luteolin, Apigenin, Quercetin, β- Sitosterol, Isoarboreal, Oleic acid, Gmelanon, Gummadiol, Palmitic acid, oleic acid,Linoleic acid, Glucose, Xylose, Mannose, Arabinose, Galactose, Rhamnose, Stigmasterol, Sitgmastanol, Compestrol, Butulinol.

***Medicinal properties :***

काश्मरी तुवरा तिक्ता वीर्योष्णा मधुरा गुरुः ।
दीपनी पाचनी मेध्या भेदिनी भ्रमशोष्यजित् ।।
दोषतृष्णाऽऽमशूलार्शोविषदाह ज्वरापहा । (भा.प्र.)

Kāśmarī tuwrā tiktā vīryosṇā madhurā guruḥ
Dīpanī pācanī medhyā bhedinī bhramaśosajịt
Doshtṛisṇāamśulāṛśovisdāh jvarāpahā." (Bhā. Pra.)

Root - Acrid, anthelmintic, appetiser, bitter, indigestion, laxative stomachic, sweet.

Flower - Astringent, bitter, cooling, sweet.

Fruits - Alterative, astringent, bitter, cooling, sweet.

***Medicinal uses :*** Root - Abdominal pain, burning sensation, fever, hallucinations, piles, urinary discharges.

Flower - Blood diseases, leprosy.

Fruits - Anaemia, leprosy, strangury, ulcers, vaginal discharges.

***Ethno- Medicinal uses :*** Bark - Stomach trouble (Jain & Borthakur, 1980).

Root - Malaria (Banerjee, 1977); stomach trouble (Tribedei *et al.*, 1982); stomach ulcer (Ganeshan *et al.*, 2006); Gout (Jadeja *et al.*, 2006).

Leaf - Cough, gonorrhoea (Sudhakar & Rolla, 1985); headache (Jain & De, 1966); wound (Sharma *et al.*, 1979); stomach ulcer (Ganeshan *et al.*, 2006).

Stem - Stomach ulcer (Ganeshan *et al.*, 2006).

## *GREWIA SUBINAEQUALIS* DC. (TILIACEAE)

***Syn.:*** *G. hainesiana* Hole.

***Vern.:*** Phalsa.

***Regional and Other Names :*** *Bengali* – Phalsa; *Gujarati* - Phalsa; *Hindi* - Dhamin; *Kannada* - Todasala, buttiyudippe; *Malayalam* – Chadicha; *Marathi* – Phalsi; *Sanskrit* - Alpasthi; *Tamil* - Palisa; *Telugu* - Jana, nallajana.

***Distribution :*** Extensively cultivated; in the wild state unknown.

***Taxonomic description :*** Shrubs or small trees, with ovate to orbicular leaves and yellow flowers in extra-axillary cymes. Petals half the length of basal glands. Nuts 1-celled.

***Flowering & Fruiting :*** November – May.

***Medicinal parts :*** Root bark, Fruits, Leaves.

***Taste :*** Sour, Bitter, Acrid.

***Chemical constituents :*** Plant contains Lupeol, Lupenon, Betulin, Friedelin, Taraxasterol, Beta-Amyrine, Grewionol, Kampferol, Amino acids, Citric acid, Vitamin-C.

***Medicinal properties :***

परूषकं कषायाम्लमामं पित्तकरं लघु ।
तत्पक्वं मधुरे पाके शीतं विष्टम्भि बृहणम् ।
हयन्तु पित्तदाहास्त्रज्वरक्षयसमीरहत् ।। (भा. प्र.)

Paruskaṃ kasāyāṃlmāmuṃ piṭtkaraṃ laghu
Taṭpakvaṃ madhure pāke sītaṃ viṣtaṃbhi bṛihṇaṃ
Hyaṇtu piṭtdāhaṣṭrajvarkshaysamīrhaṭ. (Bhā. Pra.)

Fruits - Aphrodisiac, appetiser, burning sensation, cooling, digestive, sweet, tonic.

***Medicinal uses :*** Fruits - Blood disorders, fever, heart disorders, inflammation, oedema, stomachahe.

***Ethno- Medicinal uses :*** Leaf - Boils (Tarafder, 1984).

Root - Debility (Srivastava & Varma, 1981); testicles inflammation (Tarafder & Rai chaudhuri, 1981).

Fruits - Stomachic (Das *et al.,* 1983).

## *GYMNEMA SYLVESTRE* (Retz.) R. Br. ex Sch. (ASCLEPIADACEAE)

***Syn.:*** *Asclepias geminata* Roxb.; *Periploca sylvestris* Retz.

***Vern.:*** Gurmar.

***Regional and Other Names :*** *Bengali* – Medasingi; *English* - Small Indian Ipecacuanha; *Gujarati* – Mardashingi; *Hindi* – Gurmar; *Kannada* – Sammagerasehambu; *Marathi* – Kavali; *Sanskrit* – Ajaballi, Mesasringi; *Tamil* – Adigam; *Telugu* – Podapatri.

***Distribution :*** Rare on shrubs, small trees or hedges.

***Taxonomic description :*** Much branched, twining shrubs. Leaves ovate-elliptic or oblong. Cymes greenish–yellow. Follicles solitary, straight, lanceolate, tapering.

***Flowering & Fruiting :*** February – April.

***Medicinal parts :*** Whole plant.

***Taste :*** Bitter.

***Chemical constituents :*** Leaves contains Gymnemic acid, quercitol, anthraquinon, saponin, calcium oxalate.

***Medicinal properties :***

मेषश्रृंगी रसे तिक्ता वातला श्वासकासहृत् ।
रूक्षा पाके कटुः पित्तव्रणश्लेष्माक्षिशूलनुत् ।। (भा. प्र.)

Mesasṛṇgī rase tiktā vātalā śvāsakāsahṛt
Ruksā pāke katuḥ piṭtavṛaṇaśleṣmāksiśūlanuṭ. (Bhā. Pra.)

Plant - Acrid, alterative, alexiteric anthelmintic, bitter, cooling, tonic.

Fruits - anthelmintic, alexiteric, bitter, pungent, stomachic.

***Medicinal uses :*** Plant - Asthma, biliousness, bronchitis, burning sensation, eye complaint, heart diseases, inflammation, leucoderma, ulcer.

Fruits - Bronchitis, heart diseases, ulcers, urinary discharges.

***Ethno- Medicinal uses :*** Root - Snake bite (Vishnu mittre, 1981); stomach trouble (Jain, 1965).

Seed - Diabetes (Sudhakar & Rolla, 1985); stomach trouble (Jain, 1965); gastric diseases (Maheshwari & Singh, 1987); eye diseases (Saxena, 1986).

Leaf - Diabetes (Ganesan *et al.,* 2006; Basak, 2006).

## *HALDINA CORDIFOLIA* (Roxb.) Ridsd. (RUBIACEAE)

***Syn.:*** *Adina cordifolia* (Roxb.) Hook. f. ex Bran

***Vern.:*** Haldu

***Regional and Other Names :*** *Bengali* - Dakom, Keli - Kadam; *English* – Haldu; *Gujarati* – Haldarvo; *Hindi* – Haldu; *Kannada* – Arsintega; *Malayalam* – Katampa; *Marathi* – Hedu; *Sanskrit* – Bahuphala, Haridru; *Tamil* – Kadambi; *Telugu* – Kamba.

***Distribution :*** From the Himalayas to Burma and through Central and S. India to Ceylon.

***Taxonomic description :*** Large tree with spreading branches; leaves elliptic-oblong; flowers sweet scented; fruits fleshy.

***Flowering & Fruiting :*** July – march.

***Medicinal parts :*** Root, Stem bark.

***Taste :*** Bitter.

***Chemical constituents :*** Fruit contains Flavonoids.

Leaves contain Ursolic acid, quercetin 3–O– galactoside.

Heartwood contains Adinin pigment, tannins.

Stem bark contains Alkaloids, tannins pentosan, lignins, starch.

Root bark contains Coumarin glycoside adicardin.

***Medicinal properties :***

हरिद्रुः शीतलस्तिक्तोः मंगल्यः पित्तवान्तिजित् ।
अंगकान्तिकरो बल्यो नानात्वग्दोषनाशनः ।। (रा.नि.)

Haridruḥ śitalśtiktōḥ mangalyaḥ pittavāntijit
Angkāntikarō balyō nānātvagdōsnāshnaḥ. (Raj.ni.)

Root - Astringent

Bark - Acrid, antiseptic, aphrodisiac, Bitter, Febrifuge, Pungent, tonic, vulnerary.

***Medicinal uses :*** Root - Dysentery.

Bark - Diseases of blood, inflammation, skin diseases.

***Ethnomedicinal uses :*** Bark - Fever (Jain & De, 1996); antiseptic (Mukherjee *et al.*, 1986; Basak, 2006); dysmenorrhoea (Hemadri, 1992); Stomach disease (Singh & Prakash, 1994); Skin diseases (Jha *et al.,* 1997; Miscarriages (Jadhav, 2006).

Leaves - Cold & cough (Jain & De, 1966); headache (Tribedi *et al.*, 1982).

Stem - Eye complaint (Goel *et al.*, 1984).

## *HELICTERES ISORA* L. (STERCULIACEAE)

***Vern.:*** Marod phali.

***Regional and Other Names :*** *Bengali* – Atmora; *English* - East Indian screw tree; *Gujarati* - Murdasing; *Hindi* - Bhendu, jonkaphal; *Kannada* – Yedamuri; *Malayalam* – Valmpiri; *Marathi* - Kewan; *Sanskrit* - Avartani; *Tamil* – Valamabui; *Telugu* – Gubathada.

***Distribution :*** From the Punjab and Bengal to Ceylon, Burma.

***Taxonomic description :*** Shrubs, with ovate to orbicular leaves and red flowers. Fruits cylindric, woody, beaked, with spirally coiled follicles.

***Flowering & Fruiting :*** September- February.

***Medicinal parts :*** Root, Leaf, Stem bark, seed, fruit.

***Medicinal properties :***

मुर्वा सरा गुरुः स्वादुस्तिक्ता पित्तास्त्रमेहनुत् ।
त्रिदोषतृष्णाहृद्रोगकण्डूकुष्ठजवरापहा ।। (भा.प्र.)

Mu̧rvā sarā guruḥ ṣvāduṣtiktā p̧ittāṣṭramehnu̧t
Ţridoshţrişhņāhŗdŗogkaņdūkuşthajvarāphā. (Bhā. Pra.)

Fruit - Astringent, demulcent, expectorant.

***Medicinal uses :*** Diarrhoea, dysentery, scabies, stomach affections.

***Ethno- Medicinal uses :*** Root - Asthma, cough (Singh & Pandey, 1980); abortifacient (Tribedi *et al.,* 1985); diabetes & snake bite (Roy & Chaturvedi, 1986); Sciatica (Sebastian & Bhandari, 1984); cuts (Jain *et al.,* 1973); Stomach troubles (Sharma *et al.,* 1979).

Leaf - Eczema (Ramachandran & Nair, 1981).

Seed - Dysentery & stomach trouble (Singh *et al.,* 1985).

Stem bark - Diarrhoea, dysentery (Singh *et al.,* 1985); ear complaint (Shah & Gopal, 1986).

Fruit - Colic & earache (Singh *et al.,* 1985); Cough, Cold, Fever (Mudgal & Pal, 1980); dysentery (Sharma & Malhotra, 1984; Jadeja *et al.,* 2006); stomach trouble (Maheshwari & Singh, 1984).

## ***HEMIDESMUS INDICUS*** (L.) R. Br. (ASCLEPIADCEAE)

***Syn.:*** *Asclepias pseudosarsa* var. *latifolia* Roxb.; *Periploca indica* Willd.; *P. emetica* Retz

***Vern.:*** Anantmul.

***Regional and Other Names :*** *Bengali* – Anantamul; *English* - Indian sarsaparilla; *Gujarati* - Durivel; *Hindi* - Salsa; *Kannada* – Karibandha; *Malayalam* - Narunari; *Marathi* - Antamul; *Sanskrit*- Sariva; *Tamil* - Nannari; *Telugu* – Sugandhipala.

***Distribution :*** Common in forest as undergrowth & on hedges.

***Taxonomic description :*** Glabrous, scandent undershrubs, with elliptic-oblong to linear-lanceolate leaves. Flowers dull yellow with purple eye. Follicles tapering to a point. Slightly falcate.

***Flowering & Fruiting :*** September– December.

***Medicinal parts :*** Root.

***Taste :*** Bitter, sweet.

***Chemical constituents :*** Root contains Essential oil, saponin, hemidester, hemidesmol, β- sitosterol, lupeol, tetracyclictriserpene, alcohol, resin, vasa acid, tannins.

***Medicinal properties :***

सारिवे द्वे तु मधुरे पित्तवातास्त्रनाशने ।
कण्डूकुष्ठवरहरे मेहदुर्गन्धनाशने ।। (ध. नि.)

Sārive dve tu madhure pịttvātāṣṭranāśne
Kaṇḍukuṣthavarhare menduṛgaṇdhnāśne. (Dha. Ni.)

Root - Alexiteric, antidiarrhoea, antipyretic, aphrodisiac, astringent, bitter, cooling, sweet.

***Medicinal uses :*** Root - Asthma, diarrhoea, blood diseases, burning sensation, bronchitis, dysentery, fever, itching, leprosy, leucoderma, leucorrhoea, loss of appetite, piles, skin diseases.

***Ethno- Medicinal uses :*** Stem bark - Cough, fever (Tarafder, 1986).

Fruit - Cold, cough, fever (Shah *et al.,* 1981).

Root - Stomach pain (Tribedi *et al.,* 1982); tonsil (Saxena & Vyas, 1983); diarrhoea (Tribedi *et al.,* 1982); fever (Sudhakar & Rolla, 1985); snake bite (Maheshwari *et al.,* 1986); gonorrhoea (Singh & Maheshwari, 1983; Prakash & Singh, 2006.); scorpion bite (Sharma *et al.,* 1979); dysentery (Maheshwari *et al.,* 1986); skin diseases (Sudhakar & Rolla, 1985; Deokule, 2006); pimples (Mokat & Deokule, 2006); spermatorrhoea (Prakash & Singh, 2006.); Gout (Jadeja *et al.,* 2006).

Leaf - Cough & diarrhoea (Upadhyay & Jaswant Singh, 2005).

## *HIBISCUS ROSA–SINENSIS* L. (MALVACEAE)

***Vern.:*** Gurhal.

***Regional and Other Names :*** *Bengali* – Jaba; *English* - China rose; *Gujarati* - Jasuva; *Hindi* - Gurhal; *Kannada* – Dasavala; *Malayalam* - Chemparathy; *Marathi* – Jaswand; *Sanskrit* -Japa; *Tamil* - Chedi; *Telugu* - Java pushpamu.

***Distribution :*** Cultivated in Gardens throughout India.

***Taxonomic description :*** Evergreen shrubs, with pink, red or rosy flowers.

***Flowering & Fruiting :*** November- February.

***Medicinal parts :*** Root, stem, leaves, floral buds, flowers.

***Taste :*** Acrid-Mucilagenous.

***Chemical constituents :*** Leaves contain Alkaloids, Glycosides, Reducing sugars, Sterols, Resins, Beta-sitosterol.

Root contain Mucilage, fibres.

Flowers contain Fat, calcium, phosphorous, Thiamine, riboflavin, ascorbic acid, cyanin chloride.

***Medicinal properties :***

जपा संग्राहिणी केश्या त्रिसंध्या कफवातजित् । (भा. प्र.)

Japā saṇgrāhiṇi keśhyā ṭriśaṇdhyā kaphavātajiṭ. (Bhā. Pra.)

Analgesic, antipyretic, aperient, aphrodisiac, cardiac, contraceptive, demulcent, diuretic, emolient, laxative, refrigerant.

***Medicinal uses :*** Cough, fever, anti-inflammatery, anti-bacterial, anti-fungal, antispasmodic, anti-viral, gonorrhoea, leucorrhoea.

***Ethno- Medicinal uses :*** Root-Cough (Shah, 1981; Basak, 2006); gonorr-hoea (Baruah & Sarma, 1984); venereal diseases, menorrhagia (Basak, 2006).

Flowers - Skin diseases, fever (Rao & Jamir, 1982); anti-fertility (Tarafder, 1983); contraception (Bhattacharjee *et al.,* 1980); inflammation (Baruah & Sarma, 1984); menorrhoea (Shah *et al.,* 1981); kidney diseases (Maheshwari *et al.,* 1980); fever (Deokule, 2006); asthma (Arinathan *et al.,* 2006); hair growth (Rajendran & Rengamani, 2006); leprosy (Basak, 2006); menstrual troubles (Das *et al.,* 2006).

Root - Gonorrhoea (Baruah & Sarma, 1984); cough (Shah *et al.,* 1981).

Leaf - Headache (Purkyastha & Subhan, 2006); wound, pimples (Pereira, 2006); hair growth (Ganesan *et al.,* 2006; Rajendran & Rengamani, 2006); dandruff (Gupta, 2006); leprosy (Basak, 2006).

## *HOLARRHENA PUBESCENS* (Buch.-Ham.) Wall. ex G. Don. (APOCYNACEAE)

***Syn.:*** *H. antidysenterica* (Roxb. ex f.) Wall.; *Wrightia antidysenterica* J.Grah.; *Echites antidysenterica* Heyne.

***Vern.:*** Kurchi.

***Regional and Other Names :*** *Bengali* – Kurchi; *English* - Easter tree; *Gujarati* - Dhowda; *Hindi* - Kurchi; *Kannada* – Beppale; *Malayalam* – Kadakappala; *Marathi* – Kuda; *Sanskrit* -Kutaja; *Tamil* - Kulappalai, veppalai; *Telugu* - Amkudu, kadaga.

***Distribution :*** Throughout India.

***Taxonomic description :*** Small trees, leaves ovate to elliptic. Flowers white, in terminal corymbose cymes, calyx lobes oblong-lanceolate, cilliate, corolla puberulous, follicles cylindric, often dotted with white spots.

***Flowering & Fruiting :*** February – June.

***Medicinal parts :*** Root, stem bark, seeds.

***Taste :*** Bitter.

***Chemical constituents :*** Root bark contains Holacetine, connesine alkaloids.

Seed contains Essential oil, conessine & kurchicine alkaloids.

Stem bark contains Connesine, Holarrhenine, Holarrhimine, kurchine, kurchicine, conamine, & conimine.

***Medicinal properties :***

कुटजः कटुको रूक्षो दीपनस्तुवरो हिमः ।
अर्शोऽतिसार पित्तास्त्रकफतृष्णाऽऽमकुष्ठनुत् ।। (भा.प्र.)

kuṭajaḥ katuko rukso dīpanastuvarō himaḥ
Aŗśoatisāra pịttāṣra kaphaṭṛṣṇāmaakuṣthanuṭ. (Bhā. Pra.)

Bark - Acrid, anthelmintic, bitter, pungent.

Flowers - Acrid, anthelmintic, appetiser, bitter, cooling.

Seed - Anthelmintic, appetiser, astringent, bitter, cooling, pungent.

***Medicinal uses :*** Bark - Biliousness, diarrhoea, diseases of spleen, dysentery, fever, leprosy, piles, skin diseases.

Flower - Biliousness, blood diseases, diarrhoea, leucoderma.

Seed - Biliousness, burning sensation, dysentery, erysipelas, hallucinations, leprosy, piles, skin diseases.

***Ethno- Medicinal uses :*** Flower - Worms (Baruah & Sarma, 1984).

Fruit - Stomach trouble (Maheswari *et al.,* 1986).

Leaf - Leucoderma (Baruh & Sarma, 1984); dysentery (Baruah *et al.,* 2006); Gout (Jadeja *et al.,* 2006).

Root - Fever (Shah & Gopal, 1986); stomach trouble (Sharma & Malhotra, 1984); dysentery (Mokat & Deokule, 2006).

Seed - Asthma & colic (Shah *et al.,* 1983); worms & diarrhoea & fever (Gupta, 1981); leprosy (Baruah & Sarma, 1984); jaundice (Sharma *et al.,* 1979); malaria (Deka *et al.,* 2006); diabetes (Ghosh, 2006).

Bark - Gout, snake bite, rheumatism, paralysis (Gupta, 1981); piles, dysentery, diarrhoea (Baruah & Sarma, 1984); cough, fever (Tarafder, 1986); snake bite (Sebastian & Bhandari, 1984); stomach diseases (Sharma & Malhotra, 1984); malaria (Singh & Maheshwari, 1983); cold (Srivatava & Varma, 1981); gastric diseases (Tarafder, 1983); headache (Joshi *et al.,* 1980).

## *HOLOPTELEA INTEGRIFOLIA* (Roxb.) Planch. (ULMACEAE)

***Vern.:*** Chilmil.

***Regional and Other Names :*** *English* - Indian elm; *Gujarati* - Kanjho; *Hindi* – Papri; *Kannada* - Thavasai, rasbija; *Malayalam* – Aval; *Marathi* – Papra; *Sanskrit* - Chirabilva;*Tamil* - Avali, kanji; *Telugu* - Nemali, tapasi.

***Distribution :*** Sub-himalayas, Ajmer, Bundelkhand, Bihar, Assam, Burma, W. Peninsula, Ceylon.

***Taxonomic description :*** Large trees, with ovate, pinnately veined leaves and greenish-yellow flowers. Fruits brodly oblong to suborbicular, winged, flat, notched at apex.

***Flowering & Fruiting :*** February– May.

***Medicinal parts :*** Bark.

***Taste :*** Bitter.

***Medicinal properties :***

करंजी स्तंभनी तिक्ता तुवरा कटुपाकिनी ।
वीर्योष्णा वमिपित्तार्शः कृमिकुष्ठप्रमेहजित् ।। (भा.प्र.)

Karaṇjī ṣtaṃbhanī tiktā tuwarā katupākinī
Vīryoṣaṇā vamipịttāṛshaḥ kṛumikuṣthapṛamehjịt. (Bhā. Pra.)

Bark, Leaves - Acrid, anti-inflammatory, anthelmintic, astringent, bitter, carminative, depurative, laxative, revulsive.

***Medicinal uses :*** Colic, diabetes, dyspepsia, flatulence, helminthiasis, inflammation, leprosy, rheumatism, skin diseases, vomiting.

***Ethno- Medicinal uses :*** Leaf - Skin diseases (Singh & Maheshwari, 1983); boils (Maheshwari *et al.,* 1986); scorpion sting (Singh *et al.,* 1987).

Stem bark - Ringworm (Maheshwari & Singh, 1984); ulcer (Apparanantham & Chelladurai, 1986); scabies (Mudgal & Pal, 1980); hydrocele (Goel *et al.,* 1984); bone fracture (Pal & Jain, 1989); Rheumatism, (Negi *et al.,* 1985; Kumar & Chauhan, 2005).

Seed oil - Rheumatism (Pal & Jain, 1989).

## *HYGROPHILA AURICULATA* (Sch.) Heine (ACANTHACEAE)

***Syn.:*** *Astercantha longifolia* Nees.

***Vern.:*** Talmakhana.

***Regional and Other Names :*** *Bengali* – Kuliakhara; *Gujarati* - Gokhru; *Hindi* - Gokulkanta; *Malayalam* – Vayalchulli; *Marathi* – Kolshinda; *Sanskrit* –Vajrakantaka, Kokilaksa; *Tamil* - Nirmulli; *Telugu* – Nirguviveru.

***Distribution :*** Throughout India.

***Taxonomic description :*** Hispid annuals, with thickened nodes. Leaves oblong or lanceolate, in verticils of 6 at a node. Axillary spines yellow. Flowers purple-blue, in axillary whorls. Capsules linear- oblong.

***Flowering & Fruiting :*** August– December.

***Medicinal parts :*** Roots, leaves, seeds.

***Taste :*** Bitter.

***Chemical constituents :*** Seed contains Yellow coloured oil; diastase, lipase, protease, salts of potassium, mucilage.

***Medicinal properties :***

क्षुरकः शीतलो वृष्यः स्वाद्वअम्लः पिच्छिलस्तथा ।
तिक्तो वातामशोथाश्मतृष्णाहृष्टयनिलास्त्रजित् ।। (भा.प्र.)

Ksurakaḥ śītalo vṛisyaḥ ṣvāḍvaṃlaḥ picchilaṣtathā
Tikto vātāmśothāśmtṛiṣṇāṛaṣtynilāṣtṛjiṭ. (Bhā. Pra.)

Latex - Aphrodisiac, bitter, sour, sweet.

Seed - Acrid, aphrodisiac, bitter, constipating, cooling, tonic..

***Medicinal uses :*** Latex - Abdominal trouble, anaemia, biliousness, constipation, diarrohoea, diseases of the eye, dysentery, inflammation, urinary calculi, urinary discharges.

Seed - Biliousness, blood diseases.

***Ethno- Medicinal uses :*** Leaf - Bodyache, jaundice (Chaudhuri *et al.*, 1975); joint pain (Ganeshan, *et al.*, 2006).

Root - Dropsy, rheumatism, jaundices (Singh & Pandey, 1980).

Whole plant - Malaria (Sexena & Vyas, 1983); fever (Hembrom & Goel, 2005); diabetes (Kiruba *et al.*, 2006).

## *JATROPHA CURCAS* L. (EUPHORBIACEAE)

***Vern.:*** Ratanjot.

***Regional and Other Names :*** *Bengali* - Bag harenda; *English* - Purging nut; *Gujarati* - Jamalgota; *Hindi* - Jangli arand, ratanjot; *Kannada* – Adaluha- ralu; *Malayalam* – Kattalavanakka; *Marathi* – Moglieranda; *Sanskrit* – Suklaer-anda, Akhuparnika; *Tamil*-Adalae, kattukkottai; *Telugu*-Adaviyamadamu.

***Distribution :*** Grown in India. Native of tropical America.

***Taxonomic description :*** Eglandular, soft wooded shrubs or small trees, with 3 to 5-angled ovate-cordet leaves. Flowers greenish-yellow, in corymbose cymes. Capsules globose-ovoid.

***Flowering & Fruiting :*** November – February.

***Medicinal parts :*** Root bark, leaves, stem, seeds, and latex.

***Taste :*** Acrid- sweet, pungent.

***Chemical constituents :*** Seed contains Amino acids, arachid acid, lionleic acid, myrstic acid, oleic acid, stearic acid, arabinose, glucose, rhamnose, galactose, xylose, galactouronic acid, β- sitoserol.

Stem bark contains B- Sitosterol, taraxerol.

Leaves contain Sucrose, vitexin, compesterol, triacontanol.

***Medicinal properties :***

शुक्लएरण्ड मधुरमुष्णं गुरु विनाशयेत ।।
शूलशोथकटीबस्तिशिरः पीडोदरज्वरान् ।
बृध्नश्वासकफानाहकासकुष्ठाममारुतान् ।। (भा.प्र.)

Śūkla eraṇd madhurmuṣnaṃ guru vināśyet
Śūlśothkaṭi basti ṣirahaḥ pidodarjwrāṇ
Bṛighnaśvāskaphā nāhakāskuṣthāmamārutāṇ. (Bhā. Pra.)

Fruit & seed - Anthelmintic, puragative.

Root - Astringent, carminative, digestive, emetic.

***Medicinal uses :*** Fruit & seed - Abdominal complaint, anemia, biliousness, diseases of the heart, dysentery, fistula, urinary discharges.

Latex - Styptic

Root bark - Anaemia, biliousness, cough, dysentery, fistula, gum trouble, inflammation, rheumatism, swelling, tuberculosis, ulcer.

***Ethno- Medicinal uses :*** Root - Abortifacient (Tarafder, 1983).

Seeds - Rheumatism, (Rajwar, 1983).

Stem - Tooth trouble (Gupta, 1981; Aditya *et al.,* 2006; Rajendran & Rengamani, 2006).

Latex - Burns & tooth trouble (Gupta, 1981; Jadhav, 2006); cancer (Chaghatai & Javed, 1978).

## *JATROPHA GOSSYPIFOLIA* L. (EUPHORBIACEAE)

***Vern.:*** Lal – erenda.

***Regional and Other Names :*** *Bengali* - Lal–bherenda; *Hindi* - Lal–erenda, Verenda; *Malayalam* – Simayavanakku; *Marathi* - Vilayati errand; *Tamil* – Adalai; *Telugu* - Nela amida.

***Distribution :*** A native of Brazil. Naturalized in many parts of India.

***Taxonomic description :*** Tall, erect, glandular hairy shrubs, with deep reddish-purple branches. Leaves orbicular to obovate. Palmately 3 to 5 lobed. Flowers blood-red. Capsules 3-lobed With oblong, strophiolate seeds.

***Flowering & Fruiting :*** August – March.

***Medicinal parts :*** Bark, leaves, seeds.

***Taste :*** Acrid.

***Chemical constituents :*** Seed contains Alanine, arginine, cystine, glycine, leucine, methionine, valline, linoleic acid, oleic acid, palitic acid.

Leaves contain Apigenin, vitexin, Isovitexin.

Bark contain Ceryl & myricyl alcohol, palmitone, β-sitosterol.

Root contain Jatrophalones A & B.

Plant contain Gossypibetelene.

***Medicinal properties :*** Plant - Emetic, purgative, stimulant.

Root - Astringent, carminative, digestive.

***Medicinal uses :*** Leaves - Carbuncles, eczema, Itching.

Latex & seed oil - Ulcer (Externally applied).

***Ethno- Medicinal uses :*** Seed - Body pain (Jain, 1965).

Leaf - Cold (Ramchandran & Nair, 1981); wound (Ganesan *et al.,* 2006).

Stem - Toothache (Sharma & Malhotra, 1984); pyorrhoea (Aditya *et al.,* 2006).

Latex - Peptic ulcer (Sarma *et al.,* 2006).

## *KALANCHOE PINNATA* (Lam.) Pers. (CRASSULACEAE)

***Vern.:*** Pattharchur.

***Regional and Other Names :*** *Bengali* - Patharkuchi; *English* - Sprout leaf plant; *Hindi* - Zakhm-e-hayat; *Malayalam* - Elamarunga; *Marathi* - Panphuti; *Sanskrit* – Asthibhaksha, Asmabheda; *Tamil* – Runakalli; *Telugu* – Ranapala.

***Distribution :*** A native of Tropical Africa, but naturalized throughout the tropics of the world.

***Taxonomic description :*** A perennial glabrous succulent herb. Leaves opposite-decussate, Compound, Leaflets ovate or elliptic, crenate or serrate. Flowers in large spreading panicles, pendent, pale-greenish and reddish-purple.

***Flowering & Fruiting :*** December-february.

***Medicinal parts :*** Leaves and Its juice.

***Taste :*** Bitter.

***Chemical constituents :*** Leaves contain Alkanes, Tricontane, Alpha- & Beta-Amyrin, Beta-sitosterol, Fumaric acid, Mallic acid, Citric acid, Quercetin, Kaempferol, Calcium sulphate, Calcium oxalate.

***Medicinal properties :***

अश्मभेदो हिमस्तिक्तः कषायो बस्तिशोधनः ।
भेदनो हन्ति दोषार्शोगुल्मकृच्छ्राश्महृद्रुजः ।
योनिरोगान्प्रमेहांश्च प्लीहशुलव्रणानि च ।। (भा. प्र.)

Aśmbhedo himstiktaḥ kasāyo baṣtiśodhnaḥ
Bhedno haṇti dosāṛśoguḷmkṛicchāśmhṛdṛujaḥ
Yonirōgāṇpṛamehāṇśca plīhśulvṛiṇāni ca. (Bhā. Pra.)

Leaves - Astringent, stimulant, styptic, wound healing.

***Medicinal uses :*** Leaves - Diarrhoea, disuria, dysentery, polyuria, tuberculosis, ulcers.

Leaves (Externally applied) - Bleeding, calculii, cuts, ulcers, wounds.

***Ethno- Medicinal uses :*** Leaf - Renal stone (Rao, 1981; Lal & Yadav, 1983; Hembrom & Goel, 2005; Mokat & Deokule, 2006); Influenza (Rao, 1981); Boils, Wounds (Baruah & Sarma, 1984; Sudhakar & Rolla, 1985;

Sebastian & Bhandari, 1984; Basak, 2006); Gastric complaint, Insect bite (Baruah & Sarma, 1984); cuts, wounds (Pereira, 2006; Das *et al.*, 2006); muscular sprain, swelling (Deokule, 2006); diarrhoea (Jadeja *et al.*, 2006); headache (Rao, 1981; Das *et al.*, 2006); dysentery (Jadeja *et al.*, 2006; Das *et al.*, 2006).

## *LANTANA CAMARA* L. (VERBENACEAE)

***Vern.:*** Vantulsi.

***Regional and Other Names :*** *Bengali* – Chotra; *English* - Wild sage; *Hindi* - Aripple; *Kannada* - Nata hu gida, hesike; *Malayalam* - Arippu, puchedi; *Marathi* - Chadurang, ghaneri; *Tamil* - Unnichedi, arippu; *Telugu* - Pulikampa, lantana.

***Distribution :*** Throughout India

***Taxonomic description :*** Prickly shrubs, prickles hooked. Leaves ovate or ovate-oblong. Flowers multicoloured, orange-yellow, pink or red. Drupes globose, black when ripe.

***Flowering & Fruiting :*** Throughout the Year.

***Medicinal parts :*** Whole plant.

***Taste :*** Acrid.

***Chemical constituents :*** Plant contains Essential oil.

Leaves contain Lantanine, lantadene, arachidic acid, linoleic acid, myristic acid, palmitic acid, citral, eugenol, furfural, geraniol, linalool, Alpha-amyrin, Beta- sitosterol. Lantanin acid, lantadene- A & B. verbascoses & lantanane.

***Medicinal properties :*** Antiseptic, antispasmodic, carminative, diaphoretic, laxative.

***Medicinal uses :*** Fistula, malaria, rheumatim, wounds.

***Ethno- Medicinal uses :*** Leaf - Itching, ringworm (Abraham, 1981); Tetanus (Purkayastha & Subhan, 2006); wounds (Deokule, 2006; Yesodharan & Sujana, 2006; Das *et al.*, 2006); skin diseases (Baruah *et al.*, 2006); cuts (Das *et al.*, 2006).

Flower - Bleeding gums & decaying teeth (Upadhyaya, & Jaswant Singh, 2005); Hepatitis (Dutta Choudhury & Shil, 2006).

## *LAWSONIA INERMIS* L. (LYTHRACEAE)

***Syn.:*** *L. alba* Lamk. Egyptian

***Vern.:*** Mehndi.

***Regional and Other Names :*** *Bengali* – Gotanti; *English* - Henna; *Gujarati* - Mendi; *Hindi* - Mehendi; *Kannada* – Mayilanchi; *Malayalam* – Mayilanji; *Marathi* – Mendi; *Sanskrit* - Medika; *Tamil* - Marudondri, maruthondi; *Telugu* - Gorintaku, aivanary.

***Distribution :*** Drier Parts of India.

***Taxonomic description :*** Much branched shrubs, with broadly lanceolate leaves and creamish, fragrant flowers. Capsules globose.

***Flowering & Fruiting :*** May – November.

***Medicinal parts :*** Leaves, flowers, fruits.

***Taste :*** Astringent.

***Chemical constituents :*** Leaves contain Heena dye, Hanno – tannic acid, tannin, olive green resin.

***Medicinal properties :***

हरीतकीचूर्णमरिष्टपत्रं चूतत्वचं दाड़िमपुष्पवृन्तम् ।
पत्रं च दद्यान्मदयन्तिकाया लेपोअंगरागो नरदेवयोग्यः ।। (भा.प्र.)

Harītakīchūṛnamariṣtpaṭraṃ chūttvachaṃ dārimpuṣpavṛaṇtaṃ
Paṭraṃ ca dadyāṇmadayaṇtikāyā lepoaṇgrāgo nardevyogyahaḥ.
(Bhā. Pra.)

Anti- inflammatory, antipyretic, burning sensation, diuretic, emetic, expectorant, prophylactic.

***Medicinal uses :*** Amenorrhoea, bronchitis, cough, haemorrhage, headache, leprosy, leucoderma, lumbago, scabies, skin diseases, stomatitis, ulcer.

***Ethno- Medicinal uses :*** Leaf - Burning sensation (Nagendra Prasad & Abraham, 1984; Deokule, 2006); piles (Jadhav, 2006); falling of hair (kumar & Chauhan, 2005); bodyache, muscular pain (Pereira, 2006); hair growth (Rajendran & Rengamani, 2006).

Root - Jaundice (Hemadri & Rao, 1984)

Seed - Weakness (Gupta, 1981).

## *LEUCAS CEPHALOTES* (Roth.) Spreng. (LAMIACEAE)

***Vern.:*** Dronpushpi.

***Regional and Other Names :*** *Bengali* – Barahalkasa; *English* - Tamba; *Gujarati* - Doshinokubo; *Hindi* - Guma; *Marathi* – Deokhumba; *Sanskrit* - Dronpuspi; *Telugu* – Peddatumni.

***Distribution :*** Punjab, Bengal, Assam, Himalaya, Rajputana Desert, Kathiawar, Gujrat, Madras.

***Taxonomic description :*** Erect, hispid herbs, with ovate, crenate-serrate leaves. Flowers white, in large, globose terminal heads. Calyx-tube bent, mouth produced on the upper side, villous within. Nutlets obovoid-oblong.

***Flowering & Fruiting :*** September– December.

***Medicinal parts :*** Whole plant, flowers.

***Taste :*** Acrid, bitter.

***Chemical constituents :*** Plant contains $\beta$- Sitosterol, flavonoides, glycosides, glucose, alkaloids, baballenic acid, aromatic oil.

***Medicinal properties :***

द्रोणपुष्पी गुरुः स्वादु रूक्षोष्णा वातपित्तकृत् ।
सतीक्ष्णालवणां स्वादुपाका कट्वी च भेदिनी ।
कफामकामलाशोषतमकश्वासजन्तुजित् ।। (भा. प्र.)

Dṛoṇpuṣhpī guruḥ ṣvādu ruksoṣṇā vātpiṭtakṛuṭ
Satīkṣṇālavaṇāṇ ṣvādupākā kaṭvī ca bhedinī
Kaphāmakāmalāśosatamakaśvāsajaṇtujiṭ. (Bhā. Pra.)

Anthelmintic, antiseptic, antispasmodic, appetiser, carminative, cathartic, diaphoretic, expectorant, heating, indigestible, laxative, stimulant.

***Medicinal uses :*** Asthma, bronchitis, cold, cough, dyspepsia, eruptions, eye complaint, headache, inflammation, jaundice, leucoma, paralysis, scabies, skin diseases, snake bite, swelling.

***Ethno- Medicinal uses :*** Whole plant - Cold (Sharma *et al.,* 1979); cough (Rajwar, 1983); chronic fever, nervous disorders (Narain & Singh, 2006b); malaria (Saxena & Vyas, 1981); skin diseases (Deokule, 2006); jaundice (Arinathan *et al.,* 2006; Dutta Choudhury & Shil, 2006).

Leaf - Headache (Maheshwari & Singh, J.P. 1984); constipation (Srivastava & Varma, 1981); fever, diarrhoea (Saxena & Vyas, 1981); nasal bleeding (Sarma *et al.,* 2006); scabies (Chatterjee & Bhatterjee, 2006); cold (Arinathan *et al.,* 2006); cough (Srivastava & Varma, 1981; Arinathan *et al.,* 2006); tonsilitis (Rajendran & Rengamani, 2006); stomach trouble (Das *et al.,* 2006).

Flower - Fever (Maheshwari & Singh, 1984).

Fruits - Fever (Maheshwari & Singh, 1984); malaria (Deka *et al.,* 2006).

## *LIMONIA ACIDISSIMA* L. (RUTACEAE)

***Syn.:*** *Feronia elephantum* Corr.; *F. limonia* (L.) Swing.

***Vern.:*** Kabit.

***Regional and Other Names :*** *Bengali* – Katbel; *English* - Elephant apple; *Gujarati* - Kothun; *Hindi* - Kait, kavitha; *Kannada* – Bela; *Malayalam* - Vila, vilatti; *Marathi* – Kawatha; *Sanskrit* -Kapittha; *Tamil* - Velamarum; *Telugu* – Velega.

***Distribution :*** Indigenous in S. India, Cultivated in many parts of India.

***Taxonomic description :*** Medium sized trees, armed spine. Leaves imparipinnate, with 3-9 leaflets. Flowers pentamerous, yellowish. Fruit globose, with woody rind.

***Flowering & Fruiting :*** February– May.

***Medicinal parts :*** Leaves, fruits & gum.

***Taste :*** Sour.

***Chemical constituents :*** Fruits contain Riboflavine, vitamin – C pectin, Iron, phosphorus.

Leaves contain Volatile oil. (Estragol).

***Medicinal properties :***

कपित्थमामं संग्राहि काषायं लघु लेखनम् ।
पक्वं गुरु तृषाहिक्काशमनं वातपित्तजित् ।। (भा.प्र.)

Kapiṭthmamāṃ saṇgrāhi kāsāyaṃ laghu lekhnuṃ
Pakvaṃ guru trsāhikkāśamanaṃ vāttapiṭtajiṭ. (Bhā. Pra.)

Fruit - Acrid, alexipharmic, aphrodisiac, refrigerant, sour, sweet.

Leaf - Astringent, carminative.

***Medicinal uses :*** Fruits - Asthma, biliousness, blood diseases, cough, dysentery, earache, heart diseases, hiccough, leucorrhoea, ophthalmia, throat diseases, tumour, vomiting.

Seed - Headache, heart diseases.

Leaf - Dysentery, hiccough, vomiting.

***Ethno- Medicinal uses :*** Root - Bodyache (Chaudhuri *et al.,* 1975).

Fruits - Diarrhoea (Bhalla *et al.,* 1982); tonic (Deokule, 2006); dysentery, dyspepsia (Bhalla *et al.,* 1982; Jadeja *et al.,* 2006).

Bark - Itching (Sexena & Vyas 1983).

## ***LUFFA ACUTANGULA*** var. ***AMARA*** (Roxb.) C. B. Clarke (CUCURBITACEAE)

***Vern.:*** Kadvi turoi.

***Regional and Other Names :*** *Bengali* – Titotorai; *English* - Bitter ribbed gourd; *Hindi* - Karvitori; *Kannada* – Kaduhire; *Malayalam* – Athanga; *Marathi* – Ranturai; *Sanskrit-* Kosataki; *Tamil* - Peypirkam; *Telugu* – Adavibira.

***Distribution :*** Throughout India.

***Taxonomic description :*** Extensive climbers, with 3-fid tendrils and palmetely 5 to 7- angled or lobed leaves. Flowers yellow; stamens 3. Fruits 10-angled.

***Flowering & Fruiting :*** September– November.

***Medicinal parts :*** Fruits, Seeds, Stem.

***Taste :*** Acrid, Bitter.

***Chemical constituents :*** Fruits contain Bitter alkaloids

Seed contains Oil & Albumin

***Medicinal properties :***

अत्यर्थम् कटुतीक्ष्णोष्णं गाढेष्विष्ट गदेषु च ।
कुष्ठपाण्ड्.वामय प्लीहशोथ गुल्मगरादिषु ।। (च. क.)

Aṭyaṛthaṃ katutikṣṇoṣṇaṃ gādheṣviṣtam gadesu ca
Kuṣthapāṇḍvāmayapḷihasothaguḷmagarādisu. (Ca. ka.)

Plant - Acrid, alexiteric, bitter, carminative, digestive, laxative, Pungent.

Fruits - Cathartic, emetic.

***Medicinal uses :*** Plant - Anaemia, ascites, biliousness, bronchitis, inflammation, jaundice, leucoderma, liver complaint, piles, tumor, uterine & vaginal tumours.

Fruits - Asthma, biliousness, bronchitis, fever, headache, leucoderma, urinary discharges.

Seed - Headache.

***Ethno- Medicinal uses :*** Fruits - Fever (Sharma & Malhotra, 1984); diabetes (Deokule, 2006).

## *MADHUCA INDICA* Gmel. (SAPOTACEAE)

***Syn.:*** *M. longifolia* (Koenig) Mach. Var. *latifolia* Chev.; *Bassia latifolia* Roxb.

***Vern.:*** Mahua.

***Regional and Other Names :*** *Bengali* – Mahua; *English* - Butter tree, mohwa tree; *Gujarati* - Mahuda; *Hindi* - Mahua; *Kannada* – Hippe; *Malayalam* – Poonamilupa; *Marathi* – Mahua; *Sanskrit* - Maduka; *Tamil* - Iiuppai; *Telugu* - Ippa.

***Distribution :*** Indigenous chiefly in the moist forest on the west side of India.

***Taxonomic description :*** Large, deciduous trees, with elliptic-obovate leaves and white flowers in axillary fascicles. Berries ovoid.

***Flowering & Fruiting :*** April – June.

***Medicinal parts :*** Flower, Fruit, Leaves, Bark, Seed oil.

***Taste :*** Sweet.

***Chemical constituents :*** Flowers Sugar, cellulose.

Seeds Oil.

***Medicinal properties :***

मधुकपुष्पं मधुरं शीतलं गुरु बृंहणम् ।
बलशुक्रकरं प्रोक्तं वातपित्तविनाशनम् ।
फलं शीत गुरु स्वादु शुक्रलं वातपित्तनुत् ।
अहृद्यं हन्ति तृष्णाऽस्त्रदाहश्वासक्षतक्षयान् ।। (भा.प्र.)

Madhukpuṣpaṃ madhuraṃ śitlaṃ guṛu bṛahṇaṃ
Balśukrakṛaktaṃ pṛoktaṃ vātpiṭtvinaśnaṃ
Phalaṃ śit guru ṣvādu śukṛlaṃ vātapiṭtanuṭ
Aḥrdyaṃ haṇti ṭrishṇāaṣṭradāhśvāskshatkshyāṇ. (Bhā. Pra.)

Plant - Acrid, anthemintic, aphrodisiac, bitter, cooling, flattening, sweet.

Flowers - Aphrodisiac, cooling, sweet.

Fruits - Aphrodisiac, cooling, indigestive, sweet.

Seed oil - Acrid, sweet.

***Medicinal uses :*** Plant - Biliousness, burning sensation, ulcer.

Bark - Bone fracture, leprosy, ulcer, wounds..

Flower - Biliousness, burning sensation, ear complaint, heart diseases.

Fruits - blood diseases, bronchitis, heart diseases.

Seed oil - Burning sensation.

***Ethno- Medicinal uses :*** Leaf - Burns (Negi *et al.,* 1985).

Seeds - Snake bite (Janaradhanan, 1963).

Seed oil - Pneumonia, chest pain (Tarafder & Rai Chaudhuri, 1981); heal cracks (Narain & Singh, 2006).

Flower - Bone fracture, colic (Nautiyal, 1981); scorpion bite (Maheshwasri & Singh, 1987); wound (Jadhav, 2006); rheumatism, swellings (Narain & Singh, 2006); cough (Nautiyal, 1981; Basak, 2006).

Bark - Diabetes, rheumatism (Bhalla *et al.*, 1982); itching, ulcer, gum trouble (Nautiyal, 1981); toothache (Goel *et al.,* 1984; Narain & Singh, 2006b); snake bite (Bedi, 1978); dibetes (Hembrom & Goel, 2005).

## *MANGIFERA INDICA* L. (ANACARDIACEAE)

***Vern.:*** Aam.

***Regional and Other Names :*** *Bengali* – Aam; *English* - Mango; *Gujarati* - Ambo; *Hindi* - Aam; *Kannada* - Mavinakayi, marv; *Malayalam* - Amram, cutam; *Marathi* – Amba; *Sanskrit* - Amra; *Tamil* - Ambiram, mau; *Telugu* - Mamidi, mavi.

***Distribution :*** Throughout India.

***Taxonomic description :*** Evergreen trees, with oblong-lanceolate leaves and polygamous flowers in panicles. Stamens and style solitary. Drupes yellow when ripe.

***Flowering & Fruiting :*** Febuary – June.

***Medicinal parts :*** Leaves, bark, flower, fruits, seed, gum.

***Taste :*** Sweet-sour.

***Chemical constituents :*** Mangiferin, Amino acids, Gallotanin, Gallic acid, Ethylgallati, Isoquercetin, β- Sitosterol, Riboflavin, Vitamin- C, citronellal, Deterpene, Geraniol, Limonene, mangiferol, Mangiferon, Nerol & Tannine, Vitamin- A, Ascorbic acid, protein, fat, carbohydrate, Calcium, phosphorous, Iron, fructose, sucrose, glucose.

***Medicinal properties :***

आम्रपुष्पमतीसारककफ पित्त प्रमेहनुत् ।
अस्रग्दुष्टिहरं शीतं रूचिकृद् ग्राहि वातलम् ।। (भा.प्र.)

Āmṛpuṣpmatisārak kaph pịtta pṛamehnụt
aṣrugdriṣtiharaṃ Śitaṃ ruchikṛuḍ grāhi vātlaṃ. (Bhā. Pra.)

Bark - Acrid, astringent, cooling.

Leaf - Acrid, astringent.

Flower - Appetizer, astringent, cooling.

Unripe fruits - Acrid, astringent, sour.

Ripe fruits - aphrodisiac, appetiser, astringent, cooling, sweet.

Seed - Acrid, sweet.

Seed oil - Acrid, bitter, sweet.

***Medicinal uses :*** Unripe fruits - Biliousness, blood impurities, dysentery, throat trouble, ulcer, urinary dischargers, vaginal trouble.

Flower - Biliousness, blood diseases, bronchitis, dysentery, leucorrhoea, urinary discharges.

Ripe fruit - Blood impurities, heart trouble, ulcers, urinary discharges.

Seed - Dysentery, vomiting.

Seed oil - Stomatitis.

***Ethno- Medicinal uses :*** Bark - Abortifacient, antifertility (Billore & audichya, 1978); Stomatitis (Rao & Jamir, 1982); Jaundice (Singh & Maheshwari, 1983); dysentery (Deokule, 2006; Sarma *et al.,* 2006).

Fruit - Diarrhoea (Shah, 1984); scorpion sting (Basak, 2006).

Gum - Toothache (Maheshwari & Singh, 1984); scorpion sting (Pereira, 2006).

Seed- Gastric problems, indigestion (Udayan *et al.,* 2006); diarrhoea (Kataria & Gauchan, 2006); dysentery (Kumar *et al.,* 2006; Kataria & Gauchan, 2006); stomachache (Arinathan *et al.,* 2006); ringworm (Rajendran & Rengamani, 2006).

Leaves – Diabetes (Kumar *et al.,* 2006; Ghosh, 2006); scorpion sting (Basak, 2006); diarrhoea, leucorrhoea, ulcers (Das *et al.,* 2006).

## *MELIA AZEDARACH* L. (MELIACEAE)

***Vern.:*** Bakayan.

***Regional and Other Names :*** *Bengali* – Ghoranim; *English* - Persian lilac; *Gujarati* - Barkanlimbodo; *Hindi* - Bakayan, Mahanima; *Kannada* - Turakabevu, Arebevu; *Malayalam* - Malaveppu, Simaveppu; *Marathi* – Bakananimb; *Sanskrit*- Mahanimbah; *Tamil* - Malaivempu, Tittam; *Telugu* - Kondavepa, yerri–vepa.

***Distribution :*** Throughout India.

***Taxonomic description :*** Trees, with 1 to 3-pinnate leaves and liliac flowers. Drupes 5-located, each locule 1-seeded.

***Flowering & Fruiting :*** April– May.

***Medicinal parts :*** Root, leaf, seed, flowers.

***Taste :*** Bitter.

***Chemical constituents :*** Plant contains Azaridine, Bakayanin, Bakalactone, Morgosine, Azadirone, Azadiradione, Orchinol, Ochinine, Ochinolal, Nimbolinin, Nimbolidins, Triterpene, Meliacins, Sendanins, Melianol, Melianone, Lupeol, β- stiosterol, Cataechin, Vanillin, Cinnamic acid, Cystine, Serin, Arginine, Glycine, Glutomic acid, Leucine, Threonine, Methionine.

***Medicinal properties :***

महानिम्बः स्मृतो द्रेका रम्यको विषमुष्टिकः ।
केशमुष्टिर्निम्बकश्च कार्मुको जीव इत्यापि ।। (भा.प्र.)

Mahānimbah smrto dreka ramyako visamustikah
Keśamustirnimbakaśca kārmuko jīva ityāpi. (Bhā. Pra.)

Root - Acrid, anodyne, anthelmintic, antiseptic, astringent, bitter, constpating, depurative, expectorant.

Seed - Anthelmintic, aphrodisiac, bitter, expectorant.

Flowers - Anodyne, astringent, diuratic, stomachic.

***Medicinal uses :*** Root - Amenorrhoea, asthma, burning sensation, cough, diabetes, dysenetery, fever, general debility helmintic, leprosy, leucoderma, lumbago, sciatica, skin diseases, ulcer, vomit, wound.

Leaf - Amenorrhoea, bronchitis, cardiac diseases, cough, hysteria, inflammation, leprosy, scabies.

Seed - Fever, helminthiasis, typhoid.

Flower - Dysenetery, fever.

***Ethno- Medicinal uses :*** Fruit - Skin diseases (Virjee *et al.,* 1984).

Bark - Skin diseases (Shah *et al.,* 1983); stomach trouble (Gupta, 2006).

Leaves - Rheumatism, gout (Singh & Maheshwari, 1983); antiseptic (Malhotra & Moorthy, 1973); headache (Roy & Chaturvedi, 1986); hair fall (Tiwari *et al.,* 1979); wound (Sharma *et al.,* 1979); promote conception (Apparanantham & Chelladurai, 1986); skin diseases (Puri, 1983); leprosy (Deokule, 2006); fever, malaria, skin diseases (Baruah *et al.,* 2006).

## *MIMOSA PUDICA* L. (MIMOSACEAE)

***Vern.:*** Chuimui.

***Regional and Other Names :*** *Bengali* – Lajjabati; *English*-Sensitive plant; *Gujarati*-Lajalu; *Hindi*-Lajwanti; *Kannada* -Lajja, Nacikegida; *Malayalam* –Thottalvadi; *Marathi* – Lajalu; *Sanskrit* - Lajjalu; *Tamil* - Thottasinigai; *Telugu* - Attipatti, Manugumaramu.

***Distribution :*** Throughout India.

***Taxonomic description :*** A diffuse undershrub clothed with long bristles from bulbous bases. Leaves sensitive, digitate; pinnae 1-2 pairs, leaflets 12-20 pairs. Flower heads dense, globose, long peduncled, very small. Corolla pink or purple. Stamens 4. Pods flat, slightly curved, with 3-5 segmented valves; sutures prickly and persistent.

***Medicinal parts :*** Root, leaf.

***Taste :*** Bitter – Astringent.

***Chemical constituents :*** Leaves contain Stigma sterol, Leucoanthocy-anidin, Flavon, Glycoside, Alkaloid – Mimosine.

Seed contains Mucilage & free Amino acids.

Seed oil contains Linolic acid, Oleic acid, Palmitic acid, Stearic acid & Behenic acid.

***Medicinal properties :***

लज्जालुर्हि शमीपत्रा समङ्गा जलविकर्णिका ।
रक्तपादी नमस्कारी नाम्ना खदिरकेत्यपि ।। (भा.प्र.)

Lajjālurhi śamīpatrā samangā jalakarṇikā
Raktapādī namaskārī nāmnā khadiraketyapi. (Bhā. Pra.)

Root - Acrid, antispasmodic, astringent, bitter, constipating, diuretic, emetic, febrifuge.

Leaf - Bitter, Sudorfic.

***Medicinal uses :*** Root - Asthma, burning sensation, dyspepsia, fever, fistula, inflammation, jaundice, leucoderma, small pox, ulcer.

Leaf - Conjunctivitis, fistula, hemorrhage hydrocele, scrofula, wounds.

***Ethno- Medicinal uses :*** Root - Snake bite (Saxena & Dutta, 1975); antifertility (Bhattacahrjee *et al.*, 1980); child birth (Chaudhuri *et al.*, 1983); dysentery (Saxena & Vyas, 1983); epilepsy (Lal, & Yadav, 1983); tootache, gum trouble (Molla, & Roy, 1985); fever (Kumar *et al.*, 1980); swelling (Udayan *et al.,* 2006 b); boils, scabies (Purkyastha & Subhan, 2006); piles (Deokule, 2006; Basak, 2006); fistula (Basak, 2006).

Leaf - Wound, hydrocoel (Nahendra Prasad & Abraham, 1984); Kideny problem (Bhalla *et al.*, 1982); boils (Kumar *et al.*, 1987); piles (Bhalla *et al.*, 1982; Deokule, 2006; Basak, 2006); Hepatitis (Dutta Choudhury & Shil, 2006); fistula (Basak, 2006); bone frature (Das *et al.,* 2006).

Flowers- Diarrhoea (Udayan *et al.,* 2006).

Whole plant - Loose motion (Tarafder, 1984); wounds (Udayan *et al.,* 2006); diarrhoea, dysentery (Kataria & Gauchan, 2006).

## *MIMUSOPS ELENGI* L. (SAPOTACEAE)

***Vern.:*** Morsali.

***Regional and Other Names :*** *Bengali* – Bakal; *English* - Bullet wood; *Gujarati* - Baphuli; *Hindi* - Mulsari; *Kannada* - Bakula, Pagademara; *Malayalam* - Batulam, Elanni; *Marathi* – Ovalli; *Sanskrit* - Bakulah; *Tamil* - Magulam, Magilam; *Telugu* - Vakulamu, Pogada.

***Distribution :*** W. Peninsula, southwards from the Khandala Ghat on the West and the N. Circas on the east side, Andamans.

***Taxonomic description :*** Evergreen trees, with elliptic-oblong leaves. Flowers white with pinkish tinge, axillary, solitary or fascicled; corolla 8-lobed; stamens 8. Berries elliptic-oblong.

***Flowering & Fruiting :*** April-June.

***Medicinal parts :*** Bark, flowers, fruits, seed.

***Taste :*** Bitter – acrid.

***Chemical constituents :*** Bark contains Lupeol & Xylose.

Leaves contain Myrcetin, Quercetin.

Flower contains Glucosides, Quercitol, Ursolic acid, Glucose, Triterpene, Essential oil, alcohol.

Seed contains Fatty oils, fatty acids – Palmitic acid, Stearic acid, Behenic acid, Oleic acid, Linoleic acid, Lauric acid, Myristic acid, Saponin.

Fruit contains Sugars.

***Medicinal properties :***

बकुलस्तुवरोनुष्णः कटुपाकरसो गुरुः ।
कफपित्तविषाश्वित्रक्रिमिदन्तगदापहः ।। (भा.प्र.)

Bakulaṣtuvarōṣṇaḥ anusnah katupākaraso guruḥ
kaphapịttavisāṣvitrakṛmidantagadāpahaḥ. (Bhā. Pra.)

Bark - Acrid, anthelmintic, astringent, cooling.

Flower - Acrid, anthelmintic, astringent, cooling.

Fruits - Acrid, anthelmintic, astringent, cooling.

Seed - Acrid, anthelmintic, astringent, cooling.

***Medicinal uses :*** Bark-Dental troubles, diarrhoea, dysentery, pyorrhoea, stomatitis.

Flower - Cough, headache, leucoderma, skeletol fracture, worms.

***Ethno- Medicinal uses :*** Bark - Fever (Bhalla *et al.*, 1982; Deokule, 2006).

Flower - Headache (Bhalla *et al.*, 1982).

## *MIMUSOPS HEXANDRA* (Roxb.) Dub. (SAPOTACEAE)

***Syn.:*** *Manilkara hexandra* (Roxb.) Dub.

***Vern.:*** Khirni.

***Regional and Other Names :*** *Bengali* – Khirkhejur; *English* - Rayan; *Gujarati* - Khirni; *Hindi* - Khirni; *Kannada* – Bakula; *Malayalam* – Pala; *Marathi* - Ranjan, Raini; *Sanskrit* - Kshirini; *Tamil* - Palla, Palai; *Telugu* - Pala, Manjipala.

***Distribution :*** Upper Gangetic Plain, Central & S. India.

***Taxonomic description :*** Small trees; bark rough, grey; Leaves oblong, elliptic or elliptic-ovate; flowers white, in fascicles of 2-6 and from the axils of follen leaves; fruits oblond-ellipsoid, reddish-yellow when ripe; seeds ovoid, reddish-brown, shining.

***Flowering & Fruiting :*** January – April.

***Medicinal parts :*** Fruits.

***Taste :*** Sweet.

***Medicinal properties :***

क्षीरिकायाः फलं वृष्यं बल्यं स्निग्धं हिमं गुरु ।
तृष्णा मुर्च्छामदभ्रांतिक्षयदोषत्रयास्त्रजित् ।। (भा.प्र.)

Kshīrikāyāḥ phalaṃ vṛiṣyaṃ bạlyaṃ ṣnighaṃ himaṃ guru
Tṛishṇā muṛchāmadbhṛāntikshaydoshṭrayāṣṭrajịt. (Bhā. Pra.)

Fruit - Aphrodisiac, cooling, indigestion, sweet, tonic.

***Medicinal uses :*** Fruit - Biliousness, hallucinations, heart diseases, leprosy, loss of consciousness.

***Ethno- Medicinal uses :*** Bark - Bodyache (Maheshwari *et al.,* 1986); chest pain (Malhotra & Moorthy, 1973); eye complaint (Saxena, 1986); toothache (Mokat & Deokule, 2006).

Fruit - Toothache (Mokat & Deokule, 2006).

Seed - Stomatitis (Jain, 1965).

## *MITRAGYNA PARVIFOLIA* (Roxb.) Korth. (RUBIACEAE)

***Syn.:*** *Stephegyne parvifolia* Korth.

***Vern.:*** Karam.

***Regional and Other Names :*** *Bengali* - Guli kadam; *Hindi* - Karam; *Kannada* – Kongu; *Malayalam* – Vimbu; *Marathi* – Kalam; *Sanskrit* - Kadambi; *Tamil* - China kadambu; *Telugu* - Nir kadamba.

***Distribution :*** Throughout the drier parts of india.

***Taxonomic description :*** Trees, with panduriform or ovate to suborbicular leaves. Flowers in heads, greenish-yellow, with mitriform stigma. Capsules oblong-ellipsoid, 10-ribbed.

***Flowering & Fruiting :*** August– February.

***Medicinal parts :*** Root, bark, leaves.

***Taste :*** Bitter.

***Medicinal properties :***

वितानो निचुलः कम्रो कादम्बी च करदम्भकः ।
नदीकदम्बः कटुकः प्रोक्तो नीरकदम्बकः ।। (भा. प्र.)

Vitāno niculaḥ kamṛo kādaṃbī ca karaṃbhakaḥ
Nadīkadaṃbaḥ katukaḥ p̣rokto nīrakadaṃkakaḥ. (Bhā. Pra.)

Root & bark - Acrid, bitter, febrifuge, stomachic.

Leaf - Acrid, anodyne, anti – inflammatory, bitter, depurative, febrifuge, stomachic, styptic, sweet, vulnerary.

***Medicinal uses :*** Root & bark - Colic, fever.

Leaf - Colic, dyspepsia, erysipelas, fever, flatulence, inflammation, leprosy, skin diseases, wound.

***Ethno- Medicinal uses :*** Leaf - Headache (Maheshwari *et al.,* 1986).

Bark - Contraceptive (Jain, 1963); fever (Tarafder, 1986); muscular pain (Janardhanan, 1963); stomatitis (Tarafder, 1963); toothache (Udayan *et al.,* 2006).

## *MOMORDICA CHARANTIA* L. (CUCURBITACEAE)

***Vern.:*** Karela.

***Regional and Other Names :*** *Bengali* – Baramasiya; *English* - Bitter gourd; *Gujarati* - Karelo; *Hindi* - Karela; *Kannada* – Karali; *Malayalam* – Kaypa; *Marathi* – Karli; *Sanskrit* - Karavelli; *Tamil* – Pakar; *Telugu* – Kakara.

***Distribution :*** Cultivated throughout India.

***Taxonomic description :*** Monoecious climbers, with 5 to 7-lobed leaves and yellow flowers. Bracts of male flowers born about or below the middle of peduncle. Fruits muricate-tuberculate.

***Flowering & Fruiting :*** April– October.

***Medicinal parts :*** Whole plant.

***Taste :*** Bitter.

***Chemical constituents :*** Plant contains Aromatic essential oil, Carotene, Glucoside, saponine, Momoridicine alkaloid, charantin (non nitrogenous substance).

***Medicinal properties :***

कारवेल्लं हिमं भेदि लघु तिक्तमवातलम् ।। (भा.प्र.)

Kārveḷlaṃ himuṃ bhedi laghu tiktamvātlaṃ. (Bhā. Pra.)

Root - Acrid, astringent, bitter.

Leaf - Antipyretic, bitter, emetic, purgataive.

Fruits - Acrid, anthelmintic, antidiabetic, anti- inflammatory, appetiser, bitter, depurative, digestive, febrifuge, purgative, stimulant, stomachic, thermogenic.

***Medicinal uses :*** Leaf - Burning sensation, constipation, fever, helminthiasis.

Fruits - Asthma, burning sensation, colic, constipation, cough, diabetes, fever gout, helminthiases, inflammation, leprosy, skin diseases, ulcer, wound.

Seed - Obstructions of the liver & spleen, ulcer.

***Ethno- Medicinal uses :*** Fruits - Rheumatism (Shah *et al.,* 1981); Oedema (Lal & Yadav, 1983); eczema (Shah & Joshi, 1971); dibetes (Shah *et al.,* 1981; Dwivedi *et al.*, 2005; Arinathan *et al.,* 2006); acidity (Pereira, 2006); malaria, rheumatism (Baruah *et al.,* 2006); diabetes (Ghosh, 2006); piles, leprosy (Basak, 2006).

Leaf - Malaria (Maheshwari & Singh, 1984; Baruah *et al.,* 2006; Deka *et al.,* 2006); rheumatism (Baruah *et al.,* 2006); piles, leprosy (Basak, 2006); diabetes (Das *et al.,* 2006).

## *MOMORDICA DIOICA* Roxb. ex Willd. (CUCURBITACEAE)

***Vern.:*** Kikodi.

***Regional and Other Names :*** *Bengali* - Bankarela; *English* - Spiny gourd; *Gujarati* – Kantolan; *Hindi* - Golkandra, kakra; *Malayalam* – Erimaposal; *Marathi* - Kartoli; *Sanskrit* –Karkotaki; *Tamil* – Paluppakkai; *Telugu* - Agakara, Puagakara.

***Distribution :*** Throughout India.

***Taxonomic description :*** Perinnial, dioecious climbers, with tuberous roots. Leaves ovate, entire or 3 to 5-lobed. Flowers yellow. Fruits ovoid, beaked, densely echinate with soft spines.

***Flowering & Fruiting :*** June – August.

***Medicinal parts :*** Leaves, fruits, tuber.

***Taste :*** Acrid.

***Chemical constituents :*** Plant contains Alkaloids, mangenese.

***Medicinal properties :***

कर्कोटी मल्हत्कुष्ठहृल्लासारुचिनाशिनी ।
श्वासकासज्वरान्हन्ति कटुपाका च दीपनी ।। (भा.प्र.)

Karkotī mạlhṛịt kuṣthahṛllāsārucināśinī
Śvāskāsjvarāṇhaṇti katupākā ca dīpnī. (Bhā. Pra.)

Leaf - Anthelmintic, aphrodisiac.

Fruits - Alexiteric, bitter, laxative, pungent, stomachic.

***Medicinal uses :*** Root - Headache, jaundice, urinary calculi.

Leaf - Asthma, bronchitis, fever, hiccough, piles.

Fruits - Asthma, biliousness, bronchitis, fever, heart trouble, leprosy, tumour, urinary discharges.

***Ethno- Medicinal uses :*** Root - Fever (Saxena & Vyas, 1981); piles (Puri, 1983); jaundice, bleeding piles (Deokule, 2006).

Whole plant - Injury (Sebasitian, 1984).

## *MORINGA OLEIFERA* Lam. (MORINGACEAE)

***Syn.:*** *M. pterygosperma* Gaertn.

***Vern.:*** Surajna.

***Regional and Other Names :*** *Bengali* –Sajina; *English* - Horse–radish tree; *Gujarati* - Midhosaragavo; *Hindi* - Sahijan; *Kannada* – Nuggi; *Malayalam* – Murinna; *Marathi* – Munagachajhada; *Sanskrit* - Sobhanjanah; *Tamil* - Murnakai; *Telugu* – Mulaga.

***Distribution :*** Indigenous in the sub-Himalayan tract from the Chenab to the Sarda, also in the Oudh forests. Cultivated throughout India.

***Taxonomic description :*** Small trees, with 3-pinnate leaves and creamy-yellow or white flowers without red streaks on petals. Fruits slender, 3-gonous, 9-ribbed.

***Flowering & Fruiting :*** Almost throughout the year.

***Medicinal parts :*** Root, bark, leaf, seed.

***Taste :*** Acrid.

***Chemical constituents :*** Root bark contains Alkaloids, Moringine, Moringinine, Spirochin, Ptrigospermine.

Leaf & fruits contain Vitamin- A, Nicotinic acid, Ascorbic acid, Vitamin- B, fatty acid.

Flower & fruits contain Amino acids, glucose & sucrose, Citric acid, Malic acid, Succinic acid, Fumaric acid, Oxalic acid.

***Medicinal properties :***

शोभान्जनः शिग्रुतीक्ष्णगंधकाक्षीव मोचकाः ।
तद्बीजं श्वेतमरिचं मधुशिग्रुस्तु लोहितः ।। (भा.प्र.)

Śobhāṇjanaḥ śigrutikṣṇagaṇdha kāksīva mochakāḥ
Taḍbījaṃ śvetamaricaṃ madhuśigruṣtu lọhitaḥ. (Bhā. Pra.)

Root - Acrid, anodyne, anthelmintic, anti- inflammatory, bitter, carminative, constpating, diuretic, dysentery, expectorant, stimulating, thermogenic.

Bark - Abortifacient, acrid, antifungal, bitter, thermogenic.

Leaf - Anodyne, anthelmintic, anti- inflammatory.

Seed - Acrid, anodyne, anti- inflammatory, antipyretic, bitter, Purgative.

***Medicinal uses :*** Root - Amenorrhoea, asthma, bronchitis, cough, diarrohoea, dysentery, dyspepsia, epilepsies, fever, hysteria, inflammation, otalgia, paralysis.

Bark - Ringworm

Leaf - Helminthiasis, inflammation, scurvy, tumours, wounds.

Seed - Fever, inflammation, neuralgia.

***Ethno- Medicinal uses :*** Root - Rheumatism, asthma, liver complaint (Rajwar, 1983); Paralysis, epilepsies (Karnick *et al.,* 1981); scorpion bite (Maheshwari *et al.,* 1986); snake bite, toothache (Purkayastha & Subhan, 2006); astringent (Shah, 2005).

Gum - Ear complaint (Karnich *et al.*, 1981).

Bark - Stomach trouble, fever (Nagendra Prasad & Abraham, 1984); eczema (Sexena & Vyas, 1983); typhoid (Tarafder, 1986); boils, toothache (Kumar *et al.,* 2006); diabetes (Ghosh, 2006); poisonous bites (Rajendran & Rengamani, 2006); rheumatism (Das *et al.,* 2006).

Leaf - Scurvy, wound (Shah, 1982); Stye (Kumar & Chauhan, 2005); night blindness (Deokule, 2006); diarrhoea, dysentery (Kataria & Gauchan, 2006); cough (Arinathan *et al.,* 2006); cold, diseases of liver, spleen (Basak, 2006).

Flowers- eye ailments (Pereira, 2006); cold (Arinathan *et al.,* 2006).

Pods- eye ailments (Pereira, 2006); cold, diseases of liver, spleen (Basak, 2006).

Seed- indigestion (Jadeja *et al.,* 2006); Gout (Jadeja *et al.,* 2006).

Seed oil – Gout, rheumatism (Kumar *et al.,* 2006).

## ***MORUS ALBA*** L. (MORACEAE)

***Vern.:*** Sehtut.

***Regional and Other Names :*** *Bengali* – Tut; *English* – Mulberry; *Gujarati* - Shetur; *Hindi* - Sehtut; *Kannada* – Hipnerle; *Malayalam* – Pattunulppulucceti; *Marathi* –Tut; *Sanskrit* –Tutam; *Tamil* - Kampilippuccicceti; *Telugu* – Pippalipanducettu.

***Distribution :*** Wild in the sub-Himalayan tract. Cultivated elsewhere.

***Taxonomic description :*** Deciduous tree. Leaves ovate to lanceolate, acute or accuminate, dentate serrate or lobed. Flowers monoecious. Male spikes elongated catkins. Female spikes short, ovoid. Fruits white or dark purple to red when ripe.

***Flowering & Fruiting :*** March-May.

***Medicinal parts :*** Root, leaves, fruits.

***Taste :*** Sweet.

***Chemical constituents :*** Fruits contain Carotene (Vitamin – A), Thiamine, Nicotanic acid, Riboflavin, Ascorbic acid, Calcium, Phosphorus, Iron.

***Medicinal properties :***

तुतं पक्वं गुरु स्वादु हिमं पित्तानिलापहम् ।
तदेवामंद्गुरु सरमम्लोष्णं रक्तपित्तकृत् ।। (भा.प्र.)

Tutaṃ pakvaṃ guru ṣvādu himaṃ piṭtānilāpahaṃ
Tadevāmaṃ guru saramaṃloṣnaṃ raktapiṭtakṛt. (Bhā. Pra.)

Root - Anthelmintic, astringent, purgative.

Leaf - Antibacterial, diaphoretic, emollient.

Fruits - Aphrodisiac, cooling, diuretic, anthelmintic, laxative, sour, sweet.

***Medicinal uses :*** Leaf – Hoarsness.

Fruits - Burning sensation, diarrhoea, dyspepsia, lumbago, small pox.

***Ethno- Medicinal uses :*** Root - Appetiser (Bennet, 1985).

***Latex*** - Ulcer (Kumar *et al.,* 1987).

## *MUCUNA PRURIENS* (L.) DC. (PAPILIONACEAE)

***Syn.:*** *M. prurita* Hook.

***Vern.:*** Kemuch.

***Regional and Other Names :*** *Bengali* – Akolshi; *English* - Cowhage; *Gujarati* - Kancha; *Hindi* - Kaunc; *Kannada* – Nasuganni; *Malayalam* – Corivalli; *Marathi* – Karacha; *Sanskrit* - Kapikacchuh; *Tamil* - Punaippidukkan; *Telugu* – Pilliadugu.

***Distribution :*** Punjab Plains, from the base of the Himalaya to Ceylon and Burma. Often cultivated.

***Taxonomic description :*** Twining herbs, with 3-foliolate leaves and purple flowers in racemes. Pods densely bristly slightly curved at both ends.

***Flowering & Fruiting :*** August– October.

***Medicinal parts :*** Root, leaf, seed.

***Taste :*** Sweet-bitter.

***Chemical constituents :*** Seed contains Carbohydrates, Steroids, Phenols, Glycosides, Tannin, Aluminium, Iron, Zinc, Calcium, Magnessium, Potessium.

***Medicinal properties :***

कपिकच्छुरात्मगुप्ता रिष्यप्रोक्ता च मर्कटी ।
अजहा कण्डूराध्यण्डा दुः स्पर्शा प्रावृषायणी ।।
लान्गुली शुकशिम्वी च सैव प्रोक्ता महर्षिभिः । (भा.प्र.)

Kapikacchurātmaguptā riṣyapṛoktā ca maṛkatī
Ajahā kaṇdūrāḍhyaṇdā duḥsparśā pṛāvṛsāyaṇī
Lāṇgulī śukaśiṃbī ca saiva pṛoktā mahaṛsibhiḥ. (Bhā. Pra.)

Root - Anthelmintic, aphrodisiac, bitter, diuretic, emollient, febrifuge, puragative, stimulant, sweet, thermogenic.

Leaf - Anthelmintic, aphrodisiac.

Seed - Anthelmintic, aphrodisiac..

Sour - Anthelmintic, aphrodisiac, astringent, laxative.

***Medicinal uses :*** Root - Amenorrhoea, constipation, dropsy, dysmenorrhoea, elephantiasis, fever, helminthiasis, ulcers.

Leaf - General debility, helminthiasis, inflammation, ulcers.

Sour - General deblility, gonorrhoea, sterility.

***Ethno- Medicinal uses :*** Root - Cough, (Jain & De, 1966); dysentery (Chaudhuri *et al.,* 1975); worm infection, urinary problem (Pereira, 2006); bleeding piles (Deokule, 2006).

Stem - Eye diseases (Pal, 1984).

Leaf - Blister, boils, ulcers (Goel *et al.,* 1984).

Seed - Urine complaint (Saxena & Vyas, 1986); snake bite (Rajwar, 1983); scorpion bite (Bhatt & Sabnis, 1987); skin diseases (Deokule, 2006).

## *MUKIA MADRASPATANA* (L.) Roem. (CUCURBITACEAE)

***Syn.:*** *Melothria madraspatana* (L.) Cogn.

***Vern.:*** Charpoti.

***Regional and Other Names :*** *Hindi* - Bilari; *Malayalam* – Mukkalpiram; *Sanskrit* - Trikosaki; *Tamil* - Musumusukkai; *Telugu* – Pottibudamu.

***Distribution :*** Throughout India.

***Taxonomic description :*** Annual climbers with wiry branches clothed with white hairs; Leaves orbicular, 3-5 lobed. Fruit globose, the size of pea, red when ripe; seeds 1-4 ellipsoids, turgid, white, and scabrous.

***Flowering & Fruiting :*** July – February.

***Medicinal parts :*** Whole plant.

***Taste :*** Bitter.

***Chemical constituents:***

Root contain Columbin.

***Medicinal properties :***

कटुकोशातकी त्वन्या पीतकः पीतपुष्पकः ।
क्ष्वेडा च पिच्छिलाह्वा च कृतरन्ध्रः त्रिकोशकी ।।

Katukośātakī tvaṇyā pītakaḥ pītapuṣpakaḥ
Kṣvedā ca picchilāḥvā ca kṛtaraṇdhaḥraḥ ṭrikośakī.

Anodyne, bitter, carminative, expectorant, refrigerant, sweet.

***Medicinal uses :*** Asthma, burning sensation, constipation, cough, dyspepsia, flatulence, neuralgia, ulcers, vertigo.

***Ethno- Medicinal uses :*** Seed - Diarrhoea, dibetes (Singh & Pandey, 1980).

Root - Toothache (Singh & Pandey, 1980).

## *MURRAYA KOENIGII* (L.) Spr. (RUTACEAE)

***Syn.:*** *Bergera koenigi* L.

***Vern.:*** Kadipatta.

***Regional and Other Names :*** *Bengali* – Barsunga; *English* - Curry leaf tree; *Gujarati* - Goranimb; *Hindi* - Mithanim; *Kannada* – Karibevu; *Malayalam* – Karuveppu; *Marathi* – Gandla; *Sanskrit* - Kaidarya; *Tamil* - Karuveppu; *Telugu* – Karivepaku.

***Distribution :*** Konkan, W. Ghats of Bombay to Travancore, Deccan, S. M. Country, Madras, Along the foot of the Himalaya from Kumaon to Sikkim.

***Taxonomic description :*** A small tree; Leaves imparipinnate, leaflets 11-25 alternate, obliquely ovate or rhomboid, irregularly crenate-dentate, pubescent beneath, sprinkled with black dots; flowers white, in corymbose cymes; fruits subglobose, black.

***Flowering & Fruiting :*** February – April.

***Medicinal parts :*** Root, bark, leaf.

***Taste :*** Acrid – aromatic, bitter, astringent.

***Chemical constituents :*** Plant contains Koenimb, Cadinine, Caryophyllene, Dipentene, Pinene, Phallandrene, Sabinene, Cavirol, Isosafrol, Terpineol, Lauric acid, Palmitic acid, Girinimbine, Koenidine, Koenigine, Koeningicine, Koenimbidine, Koenine, Mahanine, Mukonicine.

Leaves contain Mahanimbidine, Mahanimbine, Curryanine, Curryangine & Murrayaline.

***Medicinal properties :***

कालशाकं गरश्लेष्मशोफघ्नं दीपनं कटु ।। (द्र. गु. सं.)

Kālaśākaṃ garaśleṣmaśophaghnaṃ dīpanaṃ katu. **(Dra. Gu. Sam.)**

Root - Acrid, anodyne, anthelmintic, anti -inflammatory, antiseptic, appetising, aromatic, astringent, bitter, carminative, constipating, cooling, depurative, febrifuge, stomachic.

Bark - Acrid, anodyne, anthelmintic, anti-inflammatory, antiseptic, appetising, aromatic, astringent, bitter, carminative, constipating, cooling, depurative, febrifuge, stomachic.

Leaf - Acrid, anodyne, anthelmintic, anti -inflammatory, antiseptic, appeti-sing, aromatic, astringent, bitter, carminative, constipating, cooling, depurative, febrifuge, stomachic.

***Medicinal uses :*** Burning sensation, colic, diarrhoea, dysentery, dyspepsia, helmintic, inflammation, leprosy, leucoderma, pruritus, skin diseases, ulcer, vomit.

***Ethno- Medicinal uses :*** Leaf - Diarrhoea (Bhalla *et al.*, 1982); malaria (Rao, 1981); snake bite (Deokule, 2006); indigestion (Jadeja *et al.,* 2006); dysentery (Bhalla *et al.*, 1982; Kumar *et al.,* 2006).

Root bark- Vomitting (Rajendran & Rengamani, 2006).

## ***NELUMBO NUCIFERA*** Gaertn. (NYMPHAEACEAE)

***Syn.:*** *Nelumbium speciosum* Willd.; *Nelumbium nelumbo* Druce.

***Vern.:*** Kamal.

***Regional and Other Names :*** *Bengali* – Kombol; *English* - Indian lotus; *Gujarati* - Suriyakamal; *Hindi* - Kamal; *Kannada* – Kamala; *Malayalam* – Ventamara; *Marathi* – Kamal; *Sanskrit* -Kamala, Pankajam; *Tamil* - Tamaval; *Telugu* – Tamara.

***Distribution :*** Throughout the warmer parts in India.

***Taxonomic description :*** Perennial, rhizomatous aquatic herbs; leaves orbicular, peltate, 32-60 cm across; flowers white to pink, solitary on long axillary peduncles, sepals caducous; petals many seriate; carpels many, free, embeded in an obconical receptacle.

***Flowering & Fruiting :*** August– December.

***Medicinal parts :*** Whole plant.

***Taste :*** Sweet, slightly bitter.

***Chemical constituents :*** Rhizome contain Glucose, tannin, fat, resin, metarbin & an alkaloid – Nelumbine.

***Medicinal properties :***

कमलं शीतलं वर्ण्यं मधुरं कफपित्तजित् ।
तृष्णादाहास्त्रविस्फोट विषवीसर्पनाशनम् ।।
पद्मबीजं हिमं स्वादु कषायं तिक्तमं गुरुः। (भा. प्र.)

Kamalaṃ śītalaṃ vaṛṇyaṃ madhuraṃ kaphapịttajịt
Tṛṣṇādāhāṣraviṣphota visavīsaṛpanāśanaṃ
Paḍmabījaṃ himaṃ ṣvadu kasāyaṃ tiktakaṃ guruḥ. (Bhā. Pra)

Stem - Anthelmintic, astringent, cooling, diuratic.

Root - Bitter, cooling, diuratic, emollient.

Leaf - Bitter, cooling, diuratic, emollient.

Flowers - Refrigerant, sweet.

Fruits & Seeds - Aphrodisiac, astringent, bitter, cooling, diuratic, sweet.

***Medicinal uses :*** Stem – Helminthiasis, leprosy, skin diseases, vomit.

Root - Cough, diarrhoea, dyspepsia, leucoderma, small pox, spermatorrhoea.

Leaf - Burning sensation, fever, leprosy.

Flowers - Bronchitis, cholera, cough, diarrohea, fever.

Fruit & Seeds - Burning sensation, fever, leprosy, leucorrhoea, pruritus, vomit.

***Ethno- Medicinal uses :*** Flowers - Diarrhoea & cholera (Bhalla *et al.*, 1982).

## *NERIUM INDICUM* Mill. (APOCYNACEAE)

***Syn.:*** *N. odorum* Soland.; *N. oleander* L.

***Vern.:*** Kaner.

***Regional and Other Names :*** *Bengali* – Karabi; *English* - Indian oleander; *Gujarati* - Kanera; *Hindi* - Kaner; *Kannada* – Kanagilu; *Malayalam* – Arali; *Marathi* – Kaneri; *Sanskrit* - Karavira; *Tamil* - Arali; *Telugu* – Karaviramu.

***Distribution :*** Throughout the Greater part of India.

***Taxonomic description :*** Evergreen shrubs, with linear-elliptic leaves and white or red flowers in cymes.

***Flowering & Fruiting :*** Throughout the year.

***Medicinal parts :*** Root, leaf.

***Taste :*** Bitter.

***Chemical constituents :*** Root bark contains Neriodorin, Nerioderin, Karabin, Glucoside (Odorin).

Leaf contains Oleandrin, Neriodin, Ursolic acid, Adynerin.

Flower contains Parafin, Benzyl alcohol, Compesterol, stigma sterol, fatty acids.

***Medicinal properties :***

करवीरद्वयं तिक्तं कषायं कटुकं च तत् ।
व्रणलाघवकृन्नेत्रकोपकुष्ठ व्रणापहम् ।।
वीर्योष्णं कृमिकण्डुघ्नं भक्षितं विषवन्मतम् । (भा.प्र.)

Karavīradvayaṃ tiktaṃ kasāyaṃ katukaṃ ca taṭ
V̩ranalāghavakr̩n̩netrakopakuṣthav̩rānapahaṃ
Vīryosṇaṃ kr̩mikaṇdughnaṃ bhaksitaṃ visavaṇmataṃ.
**(Bhā.Pra.)**

Root/leaf - Acrid, anthelmintic, aphrodisiac, astringent, bitter, diuretic, febrifuge, stomachic, thermogenic.

***Medicinal uses :*** Root - Asthma, leprosy, pruritus, ulcers.

Leaf - Haemorrhoids, scabies.

***Ethno- Medicinal uses :*** Root - Abortifacient (Tarafder, 1983)

Flower - Eye diseases (Lal. & Yadav, 1983).

Latex - Eye diseases (Chaudhuri & Pal, 1976); blisters, boils (Sebastian & Bhandari, 1984); ringworm (Kumar & Chauhah, 2005).

Leaf- skin diseases, eye troubles (Das *et al.,* 2006).

## *NYCTANTHES ARBOR-TRISTIS* L. (OLEACEAE)

***Vern.:*** Harsinghar.

***Regional and Other Names :*** Bengali – Sheuli; English - Night jasmine; Gujarati - Jayaparvatai; Hindi - Harsingar; Kannada – Parijata; Malayalam – Parijatam; Marathi – Parijatak; Sanskrit -Sephali; Tamil - Pavilamalligai; Telugu – Pagadamalli.

***Distribution :*** Cultivated in many parts of India.

***Taxonomic description :*** Large shrubs, with 4-angled branches and ovate or ovate-lanceolate leaves. Flowers white to orange, fragrant. Capsule orbicular, compressed.

***Flowering & Fruiting :*** August – November.

***Medicinal parts :*** Leaf, flower, seed.

***Taste :*** Bitter.

***Chemical constituents :*** Leaf & stem contains Beta- Sitosterol, Nyctanthic acid, Iridoid, Glycosides- Arbortristosides A, B, C, D & E Glucomannan, Glycerides of Lignocetric, Linocetric, Myristic acid, Oleic acid, palmitic acid, Stearic acid (In Seeds), As tragalin, Nictoflorin, Nyctanthoside, Arborsides, Ascorbic acid, Carotene, Glucose, Fructose, Manitol.

***Medicinal properties :***

शेफालिः कटुतिक्तोष्णा रूक्षा वातक्षयापहा ।
स्याद्‌ंगसंधिवातध्नो गुदवातादिदोषनुत् ।। ( रा.नि.)

Śephāliḥ katukitoṣṇā ruksā vātaksyāpahā
Ṣyāḍaṇgasandivātaghno gudvātādidosanuṭ. (Rā. Ni.)

Leaf - Acrid, anodyne, anthelmintic, antibacterial, anti- inflammatory, bitter, depurative, digestive, diuretic, expectorant, febrifuge, laxative, thermogenic.

Flower - Astringent, bitter, carminative stomachich.

***Medicinal uses :*** Leaf - Baldness, bronchitis, constipation, cough, dyspepsia, fever, helminthiasis, inflammation, pruritus, sciatica.

Seed - Scurvy.

***Ethno- Medicinal uses :*** Flower - Skin diseases (Agrawal, 1981).

Bark - Bone fracture (Jain, 1987); Internal Injury (Jain, 1965).

Leaf - Fever (Singh & Maheshwari, 1983; Purkayastha & Subhan, 2006; Das *et al.,* 2006); Malaria (Singh, & Maheshwari, 1983; Narain & Singh, 2006b);Sciatica (Bhalla *et al.,* 1982; Hembrom & Goel, 2005 ); Rheumatism (Shah, 1982); cough (Chaudhri *et al.,* 1985); snake bite (Murthy *et al.,*1986); anthelmintic (Negi *et al.,* 1985); hiccups (Tarafder, 1983); joint diseases (Dwivedi *et al.,* 2005).

Leaf juice- fever (Deokule, 2006).

## *NYMPHAEA NAUCHALI* Burm f. (NYMPHAEACEAE)

***Syn.:*** *N.* lotus Hook. f. & Thoms non L.; *N. pubescens* Willd.; *N. rubra* Roxb. ex Salisb.

***Vern.:*** Kanval.

***Regional and Other Names :*** *Bengali* – Shaluk; *English* - Indian red water lily; *Gujarati* - Kanvel; *Hindi* - Koka; *Kannada* - Nyadale huvu; *Malayalam* – Periambal; *Marathi* – Lalakamal; *Sanskrit*- Kumuda; *Tamil* - Vellambal; *Telugu* – Allikada.

***Distribution :*** Common Throughout India in the warmer parts.

***Taxonomic description :*** Rhizomatous perennial floating herbs; leaves large, 15-20 cm across; orbicular, regularly sinuately, toothed, densely tomentose beneath, flowers white to blue about 10 cm across.

***Flowering & Fruiting :*** August – December.

***Medicinal parts :*** Rhizomes, flowers, seeds.

***Medicinal properties :***

कुमुदं पिच्छिलं स्निग्धं मधुरं ह्लादि शीतलम् ।। (भा. प्र.)

Kumudaṃ picchilaṃ ṣnigdhaṃ madhuraṃ ḥlādi śītalaṃ.
(Bhā. Pra.)

Rhizome - Bitter, cooling, sweet, tonic.

Flower - Astringent, cardiotonic.

Seed - Aphrodisiac, constipation, cooling, stomachic, sweet.

***Medicinal uses :*** Rhizome-Diarrhoea, dysentery, dyspepsia, debility, gonorrhoea.

Seed - Diarrhoea, dispepsia.

***Ethno- Medicinal uses :*** Rhizome- diarrhoea, dysentery (Kataria & Gauchan, 2006).

## *OCIMUM BASILICUM* L. (LAMIACEAE)

***Syn.:*** *O. caryophyllantum* Roxb.; *O. pilosum* Willd.; *O. anisatum; Basilicum citratum.*

***Vern.:*** Marwa.

***Regional and Other Names :*** *Bengali* – Babuitulsi; *English* - Sweet basil; *Gujarati* – Damaro; *Hindi* - Babuitulsi; *Kannada* – Kamakasturi; *Malayalam* – Ramattulasi; *Marathi* – Marva; *Sanskrit* – Barbari; *Tamil* - Tirunittru; *Telugu* – Bhutulasi.

***Distribution :*** Throughout the Greater part of India.

***Taxonomic description :*** An erect branching herb. Stem and branches green or purplish. Leaves ovate-lanceolate, entire or toothed. Flowers whitish-pink or purplish, in whorls on simple or branched racemes. Corolla glabrous or variously pubiscent. Nutlets black, pitted.

***Flowering & Fruiting :*** Throughout the year.

***Medicinal parts :*** Whole plant.

***Taste :*** Bitter, Acrid.

***Chemical constituents :*** Plant contains Ocimene, Methyl chavicol, Ceneol, Linoalool, Pinene- Camphor terpin hydrate, aromatic oil.

***Medicinal properties :***

बर्बरी कवरी तुंगी खरपुष्पाजगन्धिका ।
पर्णासस्तत्र कृष्णे तु कटिल्लक कुठेरको ।।
तत्र शुक्लेऽर्जकः प्रोक्तो वटपत्रस्ततोऽपरः । (भा.प्र.)

Baṛbarī kavarī tuṇgī kharapuṣpājagaṇdhikā
Paṛṇāsaṣtaṭra kṛsṇe tu kaṭillakakutherako
Taṭra sukleaṛjakaḥ pṛokto vatapaṭraṣtatōaparaḥ. (Bhā. Pra.)

Acrid, anodyne, anthelmintic, antibacterial, antidiarrhoeal, antiemetic, anti-inflammatory, antipyretic, antispasmodic, appetiser, aromatic, bitter, carminative, depurative, digestive, diuretic, expectorant, insecticidal, pectoral, stimulating, stomachic, thermogenic.

***Medicinal uses :*** Affections, agalactia, amenorrhoea, asthma, bronchitis, cardiac debility, colic, cough, diarrohoea, dysentery, dyspepsia, fever flatulence, helminthiasis, inflammation, leucoderma, neuralgia, otalgia, pruritus, rheumatism, ringworm, skin diseases, spasmodic, vomit.

***Ethno- Medicinal uses :*** Leaf - Ringworm (Jadhav, 2006); cough (Deokule, 2006; Baruah *et al.*, 2006; Gupta, 2006); skin diseases (Baruah *et al.*, 2006); head- ache (Arinathan *et al.*, 2006); cold, fever (Gupta, 2006); Gout (Jadeja *et al.*, 2006).

Seed - Diarrhoea (Das *et al.*, 1983; Kataria & Gauchan, 2006); dysentery (Malhotra & Moorthy, 1973; Kataria & Gauchan, 2006).

Flower - Cholera, dysentery, stomachache (Rao & Jamir, 1982); cough (Das *et al.*, 1983); headache (Srivastava & Varma, 1981); wounds (Pal & Srivastava, 1976).

## *OCIMUM CANUM* Sims. (LAMIACEAE)

***Syn.:*** *O. americanum* Aut. non L.

***Vern.:*** Jungli tulsi.

***Regional and Other Names :*** *English* - Sacred basil; *Hindi* - Jungli tulsi; *Malayalam* - Kattu ramatulasi; *Sanskrit* - Arjaka; *Tamil* - Ganjamkorai; *Telugu* – Kukkatulasi.

***Distribution :*** Plains and Lower Hills of India.

***Taxonomic description :*** Herbs or undershrubs, with ovate-lanceolate or ovate-oblong leaves. Flowers white, whorled in long racemes. Calyx-tube hirsute-annulate within. Nutlets ovoid.

***Flowering & Fruiting :*** July – December.

***Medicinal parts :*** Whole plant, leaves, seeds.

***Taste :*** Pungent, aromatic.

***Chemical constituents :*** Plant contains Glycosides, alkaloids, tannins, savonin, volatile oil, vitamin- C, carotene.

Seed contains Green coloured oil, systosterol, fatty acid, palmitic acid, oleic acid, linoleic acid, linolenic acid, mucilage.

***Medicinal properties :***

अर्जकः कटुका रूक्षा कटुपाका विदह्हितः ।
तीष्णा रूचिकरा हृदया दीपना लघवोऽहिमाः ।।
पित्तलाः कफवाताघ्नाघ्नन्ति कण्डूविषकृमनि ।। (कै. नि.)

Aṛjakaḥ katukā ruksā katupākā vidahinaḥ
Tīkṣṇā rucikarā hṛdyā dīpanā laghavo himāḥ
Pịttalāḥ kaphavātāghnāghnāṇti kaṇdūvisakṛmin. (Kai. ni.)

Demulcent, expectorant, stimulant, stomachic.

***Medicinal uses :*** Antispasmodic, asthma, diabetes, itching, malaria, ringworm, spermatorrhoes, tuberculosis, urino- genital diseases.

***Ethno- Medicinal uses :*** Leaf - Cold (Ramachandran & Nair, 1981); cough (Singh & Pandey, 1980); hysteria & epilepsy (Goel *et al.*, 1984); headache (Maheshwari *et al.*, 1986).

Seed - Ulcer (Maheshwari *et al.,* 1986); diarrhoea (Audichya *et al.,* 1983).

Root - Cough (Singh & Pandey, 1980).

## *OCIMUM GRATISSIMUM* L. (LAMIACEAE)

***Syn.:*** *O. frutescens*; *Citratum zeylanicum.*

***Vern.:*** Ramtulasi.

***Regional and Other Names :*** *Bengali* - Ram tulsi; *English* - Shrubby basil; *Gujarati* - Ramtulasi; *Hindi* - Bantulsi; *Kannada* – Nimmatulasi; *Malayalam* – Kattutrittavu; *Marathi* - Ram tulasi; *Sanskrit* – Vanababarika, Ramtulasi; *Tamil* - Elimucham tulasi; *Telugu* - Nimma tulasi.

***Distribution :*** Throughout India.

***Taxonomic description :*** Tall, strongly aromatic shrubs; leaves ovate-lanceolate, tapering at base and decurrent on the petiole; flowers white, in verticillasters; nutlets 2 mm across.

***Flowering & Fruiting :*** November– March.

***Medicinal parts :*** Whole plant, leaves, seeds.

***Taste :*** Pungent.

***Chemical constituents :*** Leaves contain Essential oil, steroids, ascorbic acid, glycosides, saponins, tannins.

***Medicinal properties :***

वनबर्वरिकाचोष्णा सुगन्धा कटुका च सा ।
पिशाचवान्तिभूतघ्नी घ्राणसन्तर्पणी परा ।।
बीजं चास्या दाहशोषनाशकं परिकीर्तितम् । (कै. नि.)

Vaṇabaṛbarikācoṣṇā sugaṇdhā katukā ca sā
Piśāca vāṇti bhūtaghnī ghṛāṇasaṇtaṛpaṇī parā
Bījaṃ cāṣyā dāhaśosanāśakaṃ parikīṛtitaṃ. (Kai. ni.)

Alexiteric, appetiser, expectorant, heating, pungent, stomachic.

***Medicinal uses :*** Colic, cough, disuria, erysipelas, fever, gonorrhoea, inflammation, leucoderma, polyuria, seminal debility, skin diseases, strangury, vomiting.

***Ethno- Medicinal uses :*** Leaf - Stomach pain (Purkayastha & Subhan, 2006).

Seed – Body heat (Mokat & Deokule, 2006).

## *OCIMUM SANCTUM* L. (LAMIACEAE)

***Syn.:*** *O. tenuiflorum* Sensu keng.; *O. hirsutum; O. tomentosum; O. virde.*

***Vern.:*** Tulsi.

***Regional and Other Names :*** *Bengali* – Kalatusi; *English* - Holy basil; *Gujarati* - Talasi; *Hindi* - Tulsi; *Kannada* – Karitulasi; *Malayalam* – Karuttatrttavu; *Marathi* – Tulasa; *Sanskrit* - Tulasi; *Tamil* - Tulaci; *Telugu* – Tulasi.

***Distribution :*** Throughout India.

***Taxonomic description :*** A much branched, woody herb. Stems and branches clothed with soft spreading purplish hairs. Leaves opposite-decussate, ovate, elliptic-oblong, entire or crenate-serrate, hairy, minutely gland dotted. Flowers purplish-pink, whorld, in racemes, often forming panicles. Nutlets broad ellipsoid, smooth.

***Flowering & Fruiting :*** Most part of the year.

***Medicinal parts :*** Whole plant.

***Taste :*** Acrid – bitter.

***Chemical constituents :*** Plant contains Essential oil Eugenol, Eugenol methyl ether, carvacrol, Caryophyllene, Ursolic acid, Apigenin, Luteolin, Ascorbic acid, Carotene, Alkaloid, Glycosides, Saponins & Tannins.

***Medicinal properties :***

तुलसी कटुकातिक्ता हृद्योष्णा दाहपित्तजित् ।
दीपनी कुष्ठहृत्श्वासपार्श्वरुक्कफवातजित् ।। (भा.प्र.)

Tulsaī katukātiktā hṛdyoṣṇā dāhapịttajịt
Dīpanī kuṣthahṛt ṣvāsapāṛṣvaruk kaphavātajịt. (Bhā. Pra.)

Acrid, aromatic, bitter, diaphoretic, digestive, diuretic, expectorant, febrifuge, stomachic, vermifuge.

***Medicinal uses :*** Asthma, bronchitis, hiccough, leucoderma, lumbago, otalgia, ringworm, skin diseases, urino- genital disorders, vomit.

***Ethno- Medicinal uses :*** Root - Fever (Nagendraprasad & Abraham, 1984).

Seed - Laxative (Dar *et al.*, 1984).

Leaf - Wound (Sharma & Malhotra, 1984); malaria (Bhalla *et al.,* 1982); fever (Shah, 1984; Purdyastha & Subhan, 2006); bronchistis (Sudhakar & Rolla, 1985); ringworm, leprosy, blood purifier (purohit *et al.,* 1985); cold (Singh & Pandey, 1980); cough (Singh & Pandey, 1980; Jadhav, 2006); abortifacient (Tarafder, 1983); stomach pain (Purkayastha & Subhan, 2005).

## *OROXYLUM INDICUM* (L.) Vent. (BIGNONIACEAE)

***Syn.:*** *Bignonia indica* L.; *Calosanthes indica* Blume.

***Vern.:*** Sona patha.

***Regional and Other Names :*** *Bengali* – Sanpatti; *English* - Indian trumpet tree; *Gujarati* - Podval; *Hindi* - Arlu; *Kannada* – Tattuna; *Malayalam* – Palakappayyani; *Marathi* – Taitu; *Sanskrit* - Syonakah; *Tamil* - Palaiyudaycci; *Telugu* – Pampana.

***Distribution :*** Throughout India.

***Taxonomic description :*** Trees, with 2 to 3-pinnate leaves. Flowers in racemes; corolla fleshy; stamens 5. Capsules elongate, compressed.

***Flowering & Fruiting :*** June – July.

***Medicinal parts :*** Root, leaf, fruit, seed.

***Taste :*** Bitter, acrid.

***Chemical constituents :*** Root bark contain Oroxylin- A bitter glycoside, acrid substance, pectin chrysie, baicalein oil.

Seed contain Tetuin, flavin, glucoside, fatty acid & Amino acids.

Bark contain Sodium salicilate.

***Medicinal properties :***

श्योनाकः दीपनः पाके कटुकस्तुवरो हिमः ।
ग्राही तिक्तोऽनिलश्लेष्मपित्तकास प्रणाशनः ।। (भा. प्र.)

Śyonakāḥ dīpanaḥ pāke katukaṣtuvaro himaḥ
Gṛāhī tiktoanilaśleṣmapittakāsa pṛaṇāśanaḥ. (Bhā. Pra.)

Root - Acrid, anodyne, anthelimintic, anti-Inflammatory, aphrodisiac, appetiser, astringent, bitter, carminative, diaphoretic, digestive, diuratic, expectorant, febrifuge, refrigerant, sweet.

Leaf - Anodyne, stomachic.

Fruits - Acrid, anthelmintic, stomachic, sweet.

***Medicinal uses :*** Root - Asthma, bronchitis, colic, cough, diarrhoea, dropsy, dysentery, dyspepsia, fever, flatulence, gout, hiccough, helmintic, inflammation, leucoderma, neuralgia, vomit.

Leaf - Flatulence, ulcers.

Fruits - Bronchitis, cardiac disorder, haemorrhoids, helminthiasis.

***Ethno- Medicinal uses :*** Seed - Piles (Saxena, 1986).

Leaf - Headache & stomachache (Baruah *et al.,* 1987).

Root - Snake bite & stomachache (Maheshwari *et al.*, 1986).

Root bark - Diarrhoea, dysentery (Nagendra prasad & Abraham, 1984); induce conception (Saxena *et al.,* 1981); miscarriage (Tarafder, 1984).

Stem bark - Body chest pain (Jian, 1965); scabies (Jain *et al.,* 1973); jaundice (Bhatt & Sabins, 1987; Mokat & Deokule, 2006); indigestion (Chetri, 2005); diarrhoea & dysentery (Purkayastha & Subhan, 2006).

## *OXALIS CORNICULATA* L. (OXALIDACEAE)

***Vern.:*** Changeri.

***Regional and Other Names :*** *Bengali* – Amrul; *English* - Indian sorrel; *Hindi* - Amrul; *Kannada* – Teltuppi; *Malayalam* – Puliyaral; *Marathi* – Ambuti; *Sanskrit* - Cangeri; *Tamil* - Puliyari; *Telugu* – Pulicintaku.

***Distribution :*** Throughout the warmer parts of India.

***Taxonomic description :*** Procumbent, apprised hairy herbs; leaves digitately 3-foliolate; leaflets subsessile cuneate or acute at base, appressed hairy; flowers yellow, 2-8 in umbellate clusters; capsule oblong, narrowed to the apex, pubescent; seeds transversely ribbed.

***Flowering & Fruiting :*** October– May.

***Medicinal parts :*** Whole plant.

***Taste :*** Sour-sweet.

***Chemical constituents :*** Plant contains Malic acid, Tarteric acid, Citric acid, Potassium Oxalate.

***Medicinal properties :***

चान्गेरी दीपनी रूच्या रूक्षोष्णा कफवातनुत् ।
पित्तलाम्ला ग्रहण्यर्शः कुष्ठातिसारनाशनी ।। (भा. प्र.)

Cāṇgerī dīpanī rucyā ruksoṣṇā kaphavātanuṭ
Pịttalāṃlā grahaṇyaṛśaḥ kuṣthātisāranāśinī. (Bhā. Pra.)

Anodyne, antibacterial, anti-inflammatory, antiseptic, astringent, carminative, constipating, diuratic, febrifuge, sour, thermogenic, vermifuge.

***Medicinal uses :*** Amenorrhoea, anemia, diarrhoea, dysentery, dysmenorrhoea, dyspepsia, fever, haemorrhages, scurvy, warts.

***Ethno- Medicinal uses :*** Leaf - Stomachache (Maheshwari & Singh, 1984); cuts (Saxena & Vyas, 1983); skin diseases (Bhalla *et al.,* 1982); eye complaint (Bhalla *et al.,* 1982; Purkayastha & Subhan, 2006); fever (Janardhanan, 1963; Deokule, 2006); wart (Purohit *et al.*, 1985); Insect bite & swelling (Shah & Joshi, 1971); appetiser (Shah *et al.*, 1981); cough (Rao, 1981); rickets (Maheshwari *et al.,* 1980); anaemia (Chhetri, 2005); diarrhoea (Kataria & Gauchan, 2006); dysentery (Saxena & Vyas, 1983; Kataria & Gauchan, 2006; Das *et al.,* 2006).

Whole plant- Scurvy (Sharma *et al.,* 2006); cuts, wounds, skin diseases, eye troubles (Das *et al.,* 2006)..

## *PANDANUS FASCICULARIS* Lamk. (PANDANACEAE)

***Syn.:*** *P. tectorius* Auct. Non Park.; *P. odoratissimus* L. f.

***Vern.:*** Kevda.

***Regional and Other Names :*** *Bengali* – Keori; *English* - Screw pine; *Gujarati* - Kewoda; *Hindi* - Kevda; *Kannada* – Kedage; *Malayalam* – Kaita; *Marathi* – Keora; *Sanskrit* - Ketaki; *Tamil* - Talal; *Telugu* – Ketaki.

***Distribution :*** Seacoast of the Indian Peninsula on both sides, Andamans.

***Taxonomic description :*** Much-branched, small trees, with aerial roots. Leaves drooping spirally arranged, ensiform, with recurved marginal spines. Flowers white, sweet scented. Fruits woody, yellowish-brown.

***Flowering & Fruiting :*** Rainy season.

***Medicinal parts :*** Root, leaf, flower.

***Taste :*** Bitter.

***Chemical constituents :*** Plant contains Physcion, Circilineol, Circilineol, n-tricontanol, β-stiosterol, stigmasterol, compesterol, Daucosterol, Palmitic acid, Stearic acid.

***Medicinal properties :***

केतकः कटुकः स्वादुर्लघुस्तिक्तः कफावहः ।। (भा.प्र.)

Katakaḥ katukaḥ svāduṛlaghuṣtiktaḥ kaphāvahaḥ. (Bhā. Pra.)

Root - Acrid, anodyne, antiseptic, aphrodisiac, bitter, carminative, depurative, emollient, febrifuge, stomachic, sweet, thermogenic.

Leaf - Acrid, aphrodisiac, bitter, depurative.

Flower - Acrid, anodyne, aphrodisiac, bitter.

***Medicinal uses :*** Root - Abortion, colic, diabetes, dyspepsia, fever, flatulence, general debility, leprosy, otalgia, sterility, ulcers, wounds.

Leaf - Leprosy, leucoderma, scabies, small pox, syphilis, tumours.

Flower - Leucoderma, otalgia, pruritus.

***Ethno- Medicinal uses :*** Root - Snake bite (Apparanantham & Chelladurai, 1986); miscarriages (Tarafder, 1984; Jadhav, 2006).

Leaf - Scabies & Syphilis (Nagendraprasad & Abraham, 1984); leprosy & skin diseases (Rao, 1981).

Oil of bracts- Rheumatic pain & headache (Deokule, 2006).

## *PAPAVER SOMNIFERUM* L. (PAPAVERACEAE)

***Vern.:*** Aphim.

***Regional and Other Names :*** *Bengali* – Pasto; *English* - Opium poppy; *Gujarati* - Aphina; *Hindi* - Aphim; *Kannada* – Gasagase; *Malayalam* – Avin; *Marathi* – Aphu; *Sanskrit* – Ahiphena, Aphuka; *Tamil* - Kasakasa; *Telugu* – Nallamandu.

***Distribution :*** Cultivated in warmer parts of India.

***Taxonomic description :*** Annual herbs, with oblong leaves and large, showy, white, purple or scarlet flowers.

***Flowering & Fruiting :*** January-April.

***Medicinal parts :*** Petals, fruits, seeds.

***Taste :*** Bitter.

***Chemical constituents :*** Root & fruits contain Reframidine, Thebain, Codeine, Cryptopine, Narcotine.

Latex contain Papavarine, Morphine, Oxalic acid.

Seed contains Trigly ceride ester, Alkaloid- Somniferine.

***Medicinal properties :***

आफूकं शोषणं ग्राहि श्लेष्मघ्नं वातपित्तलम् ।
आक्षेपशमनं निद्राजननं मदकारि च ।। (भा.प्र.)

Āphūkaṃ śosaṇaṃ grāhiṣieṣmaghnaṃ vātapiṭtalaṃ
Āksepaśanaṃ niḍrājananaṃ madakāri ca. (Bhā. Pra.)

Petals - Bitter, expectorant, sedative.

Fruits - Opium obtained from fruits is anodyne, antispasmodic, aphrodisiac, astringent, bitter, constipating, narcotic, sedative, sweet.

Seed - Aphrodisiac, constipating, sweet.

***Medicinal uses :*** Plant - Cough.

Fruits - Cough, Internal haemorrhages.

Seed - Diarrhoea, dysentery.

***Ethno- Medicinal uses :*** Latex - Abortifacient, body pain, cholera, dysentery (Rao & Jamir, 1982).

Seed- Diarrhoea, dysentery (Jadeja *et al.,* 2006).

**PERGULARIA DAEMIA** (Forsk.) Chiov. (ASCLEPIADACEAE)

**Syn.:** *P. extensa* (Jacq.) R. Br.; *Daemia extensa* (Jacq.) R. Br.

**Vern.:** Utaran.

**Regional and Other Names :** *Bengali* – Chagulabanti; *Gujarati* – Amaradudheli; *Hindi* – Utaran; *Kannada* – Juttuve; *Malayalam* – Velipparutti; *Marathi* – Mendhadudhi; *Sanskrit* – Kurutakah; *Tamil* – Velipparutti; *Telugu* – Jittupaku.

**Distribution :** Throughout India.

**Taxonomic description :** Twining undershrubs, with cordate-ovate and acuminate leaves. Flowers yellowish-green, in drooping, corymbose cymes; corolla salver- shaped. Follicles echinate.

**Flowering & Fruiting :** December – March.

**Medicinal parts :** Whole plant.

**Taste :** Bitter.

**Medicinal properties :**

उत्तसारणिका शीता कषाया व्रणशोधनी
जयेद् दद्रुमूत्रकृच्छगर्भयोनिरूजानिलान् ।। (कै.नि.)

Uṭtasārṇikā śītā kasāyā vṛinśhodhinī
Jayeḍ dadṛumuṭrakṛucchgaṛbhayonirujānilāṇ. (Kai. Ni.)

The plant is acrid, anthelmintic, antipyretic, astringent, emetic, expectorant, laxative, thermogenic.

**Medicinal uses :** Amenorrhoea, asthma, cough, dysmenorrhoea, fever, inflammation, leucoderma.

**Ethno- Medicinal uses :** Leaf - Asthma (Nagendra Prasad & Abraham, 1984); diarrhoea (Sudhakar & Rolla, 1985); rheumatism, snakebite (Sudhakar & Rolla, 1985); cough, asthma (Deokule, 2006); skin diseases (Arinathan *et al.*, 2006); body pain (Rajendran & Rengamani, 2006).

Whole plant - Asthma (Uniyal & Chauhan, 1973); leucoderma (Shah *et al.*, 1983); Gout (Jadeja *et al.*, 2006).

## *PHYLA NODIFLORA* L. (VERBENACEAE)

***Vern.:*** Jalpippali.

***Regional and Other Names :*** *English* - Purple lippia; *Gujarati* – Ratoliya; *Hindi* – Jalpapli; *Kannada* – Nelahippali; *Malayalam* – Niritippali; *Marathi* – Ratoliya; *Sanskrit* – Jalapippali; *Tamil* – Potuttalai; *Telugu* – Bokkena.

***Distribution :*** Throughout India.

***Taxonomic description :*** A creeping, much branched perennial herb; stem 4-angled; leaves obovate, serrate-dentate margin, fleshy, hairy; flowers white or pale lilac, densely arranged into axillary heads; fruits globose-oblong, dry, splitting into 1-seeded glabrous pyrenes.

***Flowering & Fruiting :*** October– April.

***Medicinal parts :*** Whole plant.

***Taste :*** Sweet, astringent.

***Chemical constituents :*** Plant contain Fructose, glucose, lactose, xylose, linoleic acid, linolenic acid.

Leaf contain Nodiflorins- A & B, nodifloridins A & B, nepetin, lippiflorin – A, nodifloretin, glucosides of β-sterol & syigma-streol.

***Medicinal properties :***

जलपिप्पलिका तिक्ता कषाया कफपित्तजित् ।
श्वासास्रविषदाहार्त्तिभ्रममूर्च्छतृषापहा ।। (भा. प्र.)

Jalapippalikā tiktā kasāyā kaphapittajịt
ṣvāsāsṛavisadāhāṛttibhṛamūṛcchatṛṣāpahā. (Bhā. Pra.)

The plant is acrid, anthelmintic, aphrodisiac, appetiser, astringent, constipating, cooling, digestive, diuratic, fever, stomachic.

***Medicinal uses :*** Asthma, bronchitis, colic, diarrhoea, dyspepsia, fever, flatulence, gonorrhoea, helmintic, ulcer.

***Ethno- Medicinal uses :*** Leaf - Boils (Malhotra & Moorthy, 1973); diuretic (Sudhakar & Rolla, 1985); boils (Kiruba *et al.*, 2006).

Whole plant - Diuretic, fever (Singh, 1945); menstrual complaint (Audichya *et al.*, 1983).

## *PHYLLANTHUS EMBLICA* L. (EUPHORBIACEAE)

***Syn.:*** *Emblica officinalis* Gaertn.

***Vern.:*** Amla.

***Regional and Other Names :*** *Bengali* – Amalki; *English* - Emblic myrobalan; *Gujarati* - Ambri; *Hindi* - Amvala; *Kannada* – Nellka; *Malayalam* – Nellikka; *Marathi* – Anvala; *Sanskrit* - Amalaki; *Tamil* - Nelli; *Telugu* – Amalakamu.

***Distribution :*** Throughout India.

***Taxonomic description :*** Large, deciduous trees, with distichous, linear-oblong leaves. Flowers greenish-yellow, in fascicles on leafless branches. Fruits globose, fleshy; seeds bony.

***Flowering & Fruiting :*** February – April.

***Medicinal parts :*** Root, bark, leaf, fruit.

***Taste :*** Sweet, sour, astringent.

***Chemical constituents :*** Fruits contain Protein, fats, fibres, Carbohydrates, Vitamin- C, Nicotinic acid, Tannins.

After drying fruits contain Gallic acid, Ellagic acid, Flavin & glucose.

Seed contains Linoleic acid, Linolenic acid, oleic acid.

***Medicinal properties :***

हन्ति वातं तदम्लत्वात् पित्तं माधुर्यम् शैत्यतः ।
कफं रूक्षकषायत्वात् फलं धात्र्यास्त्रिदोषजित् ।। (भा.प्र.)

Haṇti vātaṃ tadaṃlạtvāṭ pịttaṃ mādhuṛyaṃ śaītyataḥ
Kaphaṃ ruksakaṣāyạtvāṭ phalaṃ dhāṭryāṣṭṛidoṣajịt. (Bhā. Pra.)

Root bark - Astringent.

Fruits - Acrid, alterant, anodyne, antipyretic, aphrodisiac, astringent, bitter, carminative, cooling, digestive, diuratic, laxative, sour, stomachic, sweet.

***Medicinal uses :*** Root bark - Diarrohoea, gonorrhoea, jaundice.

Leaf - Conjunctivitis, diarrohoea, dysentery, dyspepsia, inflammation.

Fruits - Anaemia, asthama, bronchitis, colic, cough, diabetes, diarrohoea, dysentery, dyspepsia, fever, flatulence, haemorrhages, inflammation, jaundice, leprosy, leucorrhoea, menorrhagia, peptie ulcer, skin diseases.

***Ethno- Medicinal uses :*** Seed - Asthma, bronchistis (Sharma *et al.*, 1979); madness (Jain & De, 1966); menstrual trouble (Das *et al.*, 2006).

Leaf - Diabetes, dysentery (Negi *et al.*, 1985); scorpion bite, wounds (Maheswari & Singh, 1987); urine problems (Baruah *et al.*, 2006).

Fruits - Fever & indigestion (Bhalla *et al.*, 1982); headache, liver complaint (Jain & De, 1966); stomach complaint (Virjee *et al.*, 1984); eye complaint (Megoneitso & Rao, 1983); diabetes (Tiwari *et al.*, 1973; Ghosh, 2006; Das *et al.*, 2006); bronchitis (Joshi *et al.*, 1980); dysentery (Purkayastha & Subhan, 2006; Rajendran & Rengamani, 2006); inflammation of the eyes (Upadhyay & Jaswant Singh, 2005); diuretic, tonic (Deokule, 2006); flatulence (Jadeja *et al.*, 2006); urine problems (Baruah *et al.*, 2006); malaria (Deka *et al.*, 2006); anaemia (Kadavul & Jisha, 2006); Hepatitis (Dutta Choudhury & Shil, 2006); scurvy (Basak, 2006); constipation (Jain & De, 1966; Das *et al.*, 2006).

## ***PHYLLANTHUS NIRURI*** Auct. non. L. (EUPHORBIACEAE)

***Syn.:*** *P. fraternus* Webster.

***Vern.:*** Bhuiamla.

***Regional and Other Names :*** *Bengali* - Bhui amla; *English* - Country gooseberry; *Gujarati* – Bhonyaanmali; *Hindi* - Jarmla, bhonyabali; *Kannada* - Nela nelli; *Malayalam* – Kirganellia; *Marathi* – Bhuiavali; *Sanskrit* – Bhudhatri; *Tamil* – Kilanelli; *Telugu* - Nela usirika.

***Distribution :*** Throughout India.

***Taxonomic description :*** Erect, annual herbs; stem subterete or obscurely angled; leaves oblong; stipules lanceolate; male flowers 1-3 together; disc lobes 6, glandular; female flowers solitary; perianth segments oblong-spathulate; seeds brown.

***Flowering & Fruiting :*** July- November.

***Medicinal parts :*** Root, fresh leaves, whole plant.

***Taste :*** Bitter, astringent.

***Chemical constituents :*** Leaves contain Bitter, Phyllanthin & Hypophyllanthin, Alkaloid, lignas, Vitamin- C, Quercetin, Astrogaln, Querscitrin, Rutin.

Root contains Glucoflavon.

Seed contains Oil which contains- Linoleic & Linolenic acid.

***Medicinal properties :***

भूधात्री वातकृतिक्ता कषाया मधुरा हिमा ।
पिपासाकासपित्तास्त्रकफकण्डूक्षतापहा ।। (भा.प्र.)

Bhūdhāṭrī vātkṛuttiktā kasāyā madhurā himā
Pipāsākāspịttāṣṭrakaphakaṇdūkṣtāphā. (Bhā. Pra.)

Acrid, alexipharmic, bitter, cooling, diuretic, sour, sweet.

***Medicinal uses :*** Anaemia, anuria, asthama, biliousness, bronchitis, hiccough, leprosy, menorrhagia, urinary discharges.

***Ethno- Medicinal uses :*** Root - Abortifacient (Tarafder, 1983); Jaundice (Mishra & Sahu, 1984).

Leaf - Allergy (Bhalla *et al.,* 1982); gonorrhoea (Mishra & Sahu, 1984); dysentery (Banerjee, 1977); jaundice (Sharma *et al.,* 1977).

Stem - Dysentery (Mishra & Sahu, 1984).

Whole plant - Jaundice (Nagendra Prasad & Abraham, 1984); dropsy & gonorrhoea (Mishra & Sahu, 1984); ulcer, dysentery (Rajwar, 1983); boils (Maheshwari *et al.,* 1980).

## *PHYSALIS MINIMA* L. (SOLANACEAE)

***Vern.:*** Tankari.

***Regional and Other Names :*** *Bengali* – Bantepariya; *English* - Country gooseberry; *Gujarati* – Parpoti; *Hindi* - Cirpoti; *Kannada* – Guddehannu; *Malayalam* – Nottanotiyan; *Marathi* – Chroboti; *Sanskrit* – Cirapotha; *Tamil* – Tottakkali; *Telugu* – Kuppante.

***Distribution :*** Throughout India.

***Taxonomic description :*** Erect herbs, with sinuate-dentate leaves. Flowers yellow, axillary, solitary; calyx 5-angled, enclosing yellow berries.

***Flowering & Fruiting :*** August- January.

***Medicinal parts :*** Whole plant.

***Taste :*** Bitter.

***Chemical constituents :*** Plant contains Steroidal physalin A, B & C. Epaxyphysalin- B, Plamitic acid, stearic acid, Oleic acid, Linoleic acid, Hexadeconoid acid, Epoxy & Hydroxy fatty acid.

***Medicinal properties :***

टंकारी वातजित्तिक्ता श्लेष्मघ्नी दीपनी लघुः ।
शोथोदरव्यथाहन्त्री हिता पीठविसर्पिणाम् ।। (भा.प्र.)

Taṇkārī vātjịttiktā śleṣhmaghnī dīpnī laghuḥ
Śothodarvyathāhaṇtṛī hitā pīthavisaṛpiṇāṃ. (Bhā. Pra.)

Appetiser, bitter, cooling, diuretic, expectorant, luxative, sweet.

***Medicinal uses :*** Bronchitis, colic, cough, pruritus, ulcers.

***Ethno- Medicinal uses :*** Whole plant - Earache (Sudhakar & Rolla, 1985).

Leaf - Fever (Maheshwari *et al.*, 1986).

Root - Gastric diseases (Krishna & Singh, 1987).

Fruits - Diarrhoea & toothache (Chhetri, 2005).

## *PLUMBAGO ZEYLANICA* L. (PLUMBAGINACEAE)

***Vern.:*** Chitrak.

***Regional and Other Names :*** *Bengali* – Chita; *English* - Ceylon leadtwort; *Gujarati* – Chitaro; *Hindi* – Chita; *Kannada* – Chitramula; *Malayalam* – Tumbakoduveli; *Marathi* – Chitramula; *Sanskrit* – Chitrak; *Tamil* – Cithiramulam; *Telugu* - Tella chitramulamu.

***Distribution :*** Throughout India.

***Taxonomic description :*** Erect or straggling undershrubs, with ovate-lanceolate leaves and white flowers in panicled racemes. Seed oblong, smooth.

***Flowering & Fruiting :*** July- March.

***Medicinal parts :*** Root.

***Taste :*** Acrid, bitter.

***Chemical constituents :*** Plant contains Volatile oil, Chitranone, Alpha & Beta- Amyrin, Lupeol, Taraxasterol, Fructose, Glucose, Ivertase, Protease, Chloroplumbagin, Droserone, Elliptinone, Zeylanone, Zeylinone, Meritone, Catechol, Tannin, Amino acids, Plumbagic acid.

Root contains Plumgagin alkaloids.

***Medicinal properties :***

चित्रकोऽनलनामा च पाठी व्यालस्तथोषणः ।
चित्रकः कटुकः पाके वह्निकृत्पाचनो लघुः ।। (भा.प्र.)

Chiṭrakoanalanāmā ca pāṭhī vyālṣtathoshanah
Chiṭrakaḥ katukaḥ pāke vahinkṛitpāchno laghuḥ. (Bhā. Pra.)

Root bark - Alterative, anthelmintic, astringent, Bitter, carminative, dry, hot, stomachic.

***Medicinal uses :*** Root - Anaemia, bronchitis, dysentery, inflammation, itching, leucoderma, liver diseases, piles.

***Ethno- Medicinal uses :*** Root - Cough & cold (Jadhav, 2006); skin diseases (Janardhanan, 1963); hydrocele (Hembrom & Goel, 2005); piles & paralysis (Bhalla *et al.,* 1982); rheumatism, dyspepsia (Shah, 1984); abortifacient (Tarafder, 1983); wounds & abdominal pain (Joshi *et al.,* 1980); antifertility (Goel *et al.,* 1984; Purkayastha & Subhan, 2006); diarrhoea

(Sharma *et al.*, 1979; Kataria & Gauchan, 2006); dysentery (Sebastian & Bhandari, 1984; Kataria & Gauchan, 2006); eczema (Shah *et al.*, 1983); Muscular pain (Jain, 1965); fever (Gupta, 1981); headache (Venugopal & Reghunathan, 1982); leucoderma (Bhatt & Sabnis, 1987); skin diseaes (Deokule, 2006).

Root bark- Constipation (Kadavul & Jisha, 2006).

Fruit- Gout (Jadeja *et al.*, 2006).

## *PONGAMIA PINNATA* (L.) Pierre (PAPILIONACEAE)

***Vern.:*** Karanj.

***Regional and Other Names :*** *Bengali* – Karanja; *English* - Indian beech; *Gujarati* – Kanaji; *Hindi* - Karanj; *Kannada* – Hongemara; *Malayalam* – Ponnu; *Marathi* – Karanja; *Sanskrit* - Karanjah; *Tamil* - Ponka; *Telugu* – Kanugacettu.

***Distribution :*** Throughout India.

***Taxonomic description :*** Trees, with imparipinnate leaves. Flowers pinkish-white. Pods obliquely oblong, woody, indehiscent, glabrous.

***Flowering & Fruiting :*** February – May.

***Medicinal parts :*** Root, bark, leaf, flower, seed oil.

***Taste :*** Bitter.

***Chemical constituents :*** Seed contains Bitter fatty acid oil, Pongamol (Hongay oil) and protein.

Plant contains Alkaloids, tannins, sugar, resin, fatty oil (Karanjin).

***Medicinal properties :***

करंजः कटुकस्तीक्ष्णो वीर्योष्णो योनिदोषहृत् ।
कुष्ठोदावर्तगुल्मार्शोव्रणकृमिकफापहः ।। (भा.प्र.)

Karaṇjaḥ katukaṣtīkṣṇo vīryoṣṇa yonidoṣahṛt
Kuṣthodāvaṛtaguḷmārsovṛaṇakṛmikaphāpahaḥ. (Bhā. Pra.)

Root bark - Acrid, alexeteric, anthelmintic, bitter.

Leaf - Anthelmintic, digestive, laxative.

Seed - Anthelmintic, acrid, bitter, depurative.

Seed oil - Anthelmintic, depurative, styptic.

***Medicinal uses :*** Root - Gonorrhoea.

Bark - Beriberi, haemorrhoids, ulcers.

Leaf - Cough, diarrohoea, dyspepsia, flatulence, gonorrhoea, leprosy.

Seed - Anaemia, fevers, hydrocele, inflammation, lumbago, otalgia.

Seed oil - Leprosy, lumbago, ulcers, scabies.

***Ethno- Medicinal uses :*** Leaf - Diabetes (Bhalla *et al.,* 1982); whooping cough (Ganeshan *et al.,* 2006); Gout (Jadeja *et al.,* 2006).

Fruit - Cough (Kumar & Chauhan, 2005).

Root - Fistula, skin diaseses, ulcer (Bhalla *et al.*, 1982); sore (Jain & Sahu, 1993).

Seed - Whooping cough, bronchitis (Bhalla *et al.,* 1982); skin diseases (Shah, 1984; Pereira, 2006); Joint pain (Saxena, & Vyas, 1983); ear complaint (Shah *et al.,* 1983); rheumatism (Saxena & Vyas, 1986); chest pain (Roy & Chaturvedi, 1986); burns (Jain, 1965).

Seed oil - Eczema, Itch (Jain, 1965); cold (Jain, & De, 1966); scabies (Janardhanan, 1963); fever (Tarafder, 1986); rheumatims (Kumar *et al.,* 2006).

## *PORTULACA OLERACEA* L. (PORTULACACEAE)

***Vern.:*** Badi loni.

***Regional and Other Names :*** *Bengali* – Kulfi; *English* - Common purslane; *Gujarati* – Loni; *Hindi* - Badi loni; *Kannada* – Budagora; *Malayalam* – Koluppa; *Marathi* – Bhuigholi; *Sanskrit* - Brhat loni; *Tamil* – Karikkirai; *Telugu* – Peddapavilakura.

***Distribution :*** All over India.

***Taxonomic description :*** Diffuse herbs, with succulent, obovate-spathulate leaves and yellow flowers. Capsules obovoid.

***Flowering & Fruiting :*** Throughout the year.

***Medicinal parts :*** Stem, leaf, seeds.

***Taste :*** Acrid, Acidic.

***Chemical constituents :*** Plant contains Crude fibres, carbohydrates, calcium, magnesium, phasphorus, Iron, potassium, copper, sulphur, & chlorine.

***Medicinal properties :***

घोटिकाऽम्ला सराचोष्णा वातकृत्कफपित्तहृत् ।। (भा.प्र.)

Ghotikāamlā sarāchoṣṇā vātkṛutkaphpitthṛit. (Bhā. Pra.)

Stem & leaves - Alexeteric, alterant, antibacterial, bitter, diuretic, salty, sour, stomchic, thermogenic.

Seed - Antidysenteric, cooling, diuretic.

***Medicinal uses :*** Stem & leaves - Diabetes, dysenetery, dyspepsia, gonorrhoea, inflammation, jaundice, otalgia, scurvy, skin diseases, tumours, vomit.

Seed - Burns, dysentery.

***Ethno- Medicinal uses :*** Whole plant - Burning sensation (Das *et al.*, 1983); headache (Shah *et al.*, 1981); blood purifier (Rao & Neogi, 1980).

Leaf - Fever & headache (Shash *et al.*, 1983); stomach trouble (Purkayastha & Subhan, 2006); Injury (Jadhav, 2006); burns, inflammation (Subhangi & Patil, D.A., 2006).

Seed- Dysentery (Jadeja *et al.*, 2006).

## *PORTULACA QUADRIFIDA* L. (PORTULACACEAE)

***Vern.:*** Choti luni.

***Regional and Other Names :*** *Bengali* – Chhotaluniya; *Gujarati* – Jhiniluni; *Hindi* – Chotaluniya; *Marathi* – Ranghol; *Sanskrit* – Laghughonika; *Tamil* – Passalaikkirai; *Telugu* - Kura, pavili.

***Distribution :*** Throughout the warmer parts of India.

***Taxonomic description :*** A small diffuse or prostrate annual. Stems filiform, glabrous, rooting at the nodes. Leaves sub-sessile, flat, and succulent. Flowers yellow, surrounded by a 4-5-leaved involucre and long silky hairs. Capsules conical.

***Flowering & Fruiting :*** August – November.

***Medicinal parts :*** Leaves, stem.

***Taste :*** Acidic, saltish.

***Chemical constituents :*** Plant contains Starch, Saponin, Tannin, lignin, mucilage.

***Medicinal properties :***

लोणी रूक्षा स्मृता गुर्वी वातश्लेष्महरी पटुः ।
अर्शोघ्नी दीपनी चाम्ला मन्दाग्निविषनाशिनी ।। (भा.प्र.)

Loṇī rukshā ṣmṛitā guṛvī vātśleṣhmharī paṭuhaḥ
Ạrśoghnī dīpnī cāṃlā maṇdāgniviṣhnāśhinī. (Bhā. Pra.)

Alterative, bitter, carminative, diuretic, laxative, sour, stomachic, vermifuge.

***Medicinal uses :*** Asthma, cough, eye diseases, fever, headache, inflammation, piles, skin diseases, ulcers, urinary discharges.

***Ethno- Medicinal uses :*** Whole plant - Skin diseases (Shekhawat & Anand, 1984).

Leaf - Ulcer & burns (Shah, 1984).

## *PSIDIUM GUAJAVA* L. (MYRTACEAE)

***Vern.:*** Jaamphaal.

***Regional and Other Names :*** *Bengali* – Goaachhi; *English* - Common guava; *Gujarati* - Jamrud; *Hindi* - Amrud; *Kannada* – Balehannu; *Malayalam* - Pera, koyya; *Marathi* – Jamba; *Sanskrit* - Perukah; *Tamil* - Koyya; *Telugu* – Jampandu.

***Distribution :*** Throughout India.

***Taxonomic description :*** A erect, woody shrub or small tree with reddish-brown wood and scaly, brownish bark.Leaves oblong or oblong-elliptic. Flowers on axillary peduncles of 1-3-flowered. Fruits berries.

***Flowering & Fruiting :*** July-September.

***Medicinal parts :*** Leaves, flower, fruits, root.

***Taste :*** Sweet, astringent.

***Chemical constituents :*** All parts contain Leucocyanidin, Querecetin, β- sitosterol.

Flower & leaves contain Ellagic acid, β- Caryophyllene, Guaijaverin, Umonene, Pinene, Niacin, Vitamin- C & B, Cadinene, β- Cophaene, Curcumene, Farnesene, Citric, Cinnamic, Malic, Oxalic, Quinic, Succinic, Tartaric acids, Riboflavin, Thiamine, Carotene, Tannins, sucrose, Glucose, Fructose, Rhamnose, Starch & Amino acids.

***Medicinal properties :***

पेरूकं तुवरं प्रोक्तं स्वादवम्लं कफकारकम् ।
सुकललं वातपित्तघ्न शीतलं च रसं मतम् ।। (भा.प्र.)

Perukaṃ tuvaraṃ p̣roktaṃ ṣvādvaṃlaṃ kaphakārakaṃ
Suklalaṃ vātapịttaghnam śitalaṃ ca rasaṃ mataṃ. (Bhā. Pra.)

Root - Anthelmintic, astringent, constipating.

Leave - Anodyne, antispasmodic, astringent, febrifuge.

Flower - Cooling, laxative

Fruits - Aphrodisiac, astringent, cooling, laxative, sour, sweet.

***Medicinal uses :*** Root - Diarrhoea, dysentery, haemorrhages, ulcers, vomit.

Leaves - Cholera, diarrhoea, epilepsy, ulcers, vomit, wounds.

Flower - Bronchitis, colic.

Fruits - Colic, diarrhoea, dyspepsia, dysentery, general debility.

***Ethno- Medicinal uses :*** Leaf - Blisters in mouth, cold, cough (Maheshwari & Singh, 1984); dysentery (Singh & Maheshwari, 1983; Krishna & Singh, 1987; Maheshwari *et al.,* 1986; Basak, 2006); gonorrhoea (Tarafder, 1984); fever & headache (Rao & Jamir, 1982); menstrual diseases (Bhattacharjee *et al.,* 1980); cold (Udayan *et al.,* 2006 b); cholera (Upadhyay & Jaswant Singh, 2005); fever (Deokule, 2006; Udayan *et al.,* 2006 b); diarrhoea (Upadhyay & Jaswant Singh, 2005; Sarma *et al.,* 2006); toothache (Aditya *et al.,* 2006).

Fruits - Jaundice (Tiwari *et al.,* 1979); constipation (Arinathan *et al.,* 2006); dysentery (Gupta, 2006).

Bark - Stomachache (Rao & Jamir, 1982); diarrhoea, leucorrhoea (Das *et al.*, 2006).

## *PSORALEA CORYLIFOLIA* L. (PAPILIONACEAE)

***Vern.:*** Bauchi.

***Regional and Other Names :*** *Bengali* – Bavachi; *English* - Babchi seeds; *Gujarati* - Babchi; *Hindi* - Bakuchi; *Kannada* – Somaraji; *Malayalam* – Karkokil; *Marathi* – Babachi; *Sanskrit* - Bakuci; *Tamil* - Karpogam; *Telugu* – Bavancalu.

***Distribution :*** Throughout India.

***Taxonomic description :*** Annual herbs, with simple, suborbicular, gland-dotted leaves. Flowers bluish-purple, packed in upper part of axillary racemes. Pods oblong, 1-seeded.

***Flowering & Fruiting :*** August – November.

***Medicinal parts :*** Seeds.

***Taste :*** Bitter, acrid.

***Chemical constituents :*** Seed contains Oil which contain Bakuchi Phenol (This contains Psoralane & Psoradin active Compound), Resin, Volatile oil.

***Medicinal properties :***

बाकुची मधुरा तिक्ता कटुपाका रसायनी।
रूक्षा हृदया श्वासकुष्ठमेहज्वर कृमि प्रणुत् ।। (भा.प्र.)

Bākuchī madhurā tiktā katupākā rasāyanī
Ruksyā hṛidyā ṣvāsakuṣthamehajvara kṛimi pṛanuṭ. (Bhā. Pra.)

Seeds - Acrid, anthelmintic, aphrodisiac, bitter, diaphoretic, diuretic, laxative, stimulating, tomachic.

***Medicinal uses :*** Leprosy, leucoderma, ulcers.

***Ethno- Medicinal uses :*** Seed - Itching & leprosy (Saxena & Vyas, 1983); skin diseases (Malhotra & Moorthy, 1973); leucoderma (Singh & Pandey, 1980); psoriasis (Saxena & Vyas, 1986).

Seed oil- Leprosy (Deokule, 2006).

## *PTEROCARPUS MARSUPIUM* Roxb. (PAPILIONACEAE)

***Vern.:*** Bijasal.

***Regional and Other Names :*** *Bengali* – Pitsal; *English* - Indian kino tree; *Gujarati* - Bia; *Hindi* - Bijasal; *Kannada* – Hannemara; *Malayalam* – Venna; *Marathi* – Asan; *Sanskrit*- Bijaka; *Tamil* - Vengai; *Telugu* – Asana.

***Distribution :*** W. Peninsula and S. India.

***Taxonomic description :*** Deciduous trees, with crooked stem and spreading branches. Leaves elliptic or ovate. Flowers pale-yellow, in fusco-paniculate racemes. Pods suborbicular, winged, glabrous.

***Flowering & Fruiting :*** June – February.

***Medicinal parts :*** Heartwood, leaf, flower, gum.

***Taste :*** Bitter-acrid.

***Chemical constituents :*** Plant contains Protein, pentosan, mucilage, pterosupin, pseudobaptigenin, liquiritigenin, garbanzol, β-cudesmol, pterostil-bene, marsupol, carpusin, propterol, marrsupinol, parsupin, oleanolic acid, tannins & kinotanic acid.

***Medicinal properties :***

बीजकः पीतसारश्च पीतशालक इत्यपि ।
बन्धूकपुष्पः प्रियकः सर्जकश्चासनः स्मृतः ।। (भा. प्र.)

Bījakaḥ pītasāraśca pītaśālaka ityapi
Bandhūkapuṣpaḥ priyakaḥ sarjakaścāsanaḥ smṛtaḥ. (Bhā. Pra.)

Heartwood - Acrid, alterant, anodyne, anthelmintic, anti-inflammatory, astringent, bitter, constipating, cooling, depurative, rejuvenating, union-Promoter.

Flowers - Appetiser, bitter, cooling, febrifuge, sweet.

Gum - Anthelmintic, antipyretic, bitter, styptic.

***Medicinal uses :*** Heartwood - Asthma, bronchitis, diabetes, diarrhoea, dysentery, cough, elephantiasis, fractures, gout, haemorrhages, inflammation, leprosy, leucoderma, skin diseases.

Leaf - Boils, skin diseases, sores.

Flowers - Fevers.

Gum - Boils, diarrhoea, fever, helmintic, psoriasis, ulcers, wounds.

***Ethno- Medicinal uses :*** Wood - Asthma (Maheahwari *et al.,* 1986); body pain & diabetes (Singh & Maheshwari, 1983); chest pain (Maheshwari *et al.*, 1980); cardiac problems (Udayan *et al.,* 2006).

Leaf - Boils (Janardhanan, 1963); sores (Sudhakar & Rolla, 1985).

Resin – Diarrhoea (Roy & Chaturvedi, 1986; Mokat & Deokule, 2006); toothache (Roy & Chaturvedi, 1986).

Bark - Dysentery (Jain, & De, 1966); toothache (Sudhakar & Rolla, 1985); chest pain (Udayan *et al.,* 2006); diarrhoea (Mokat & Deokule, 2006).

Flowers - Fever (Sudhakar & Rolla, 1985).

## ***RAUVOLFIA SERPENTINA*** (L.) Benth. ex Kurz. (APOCYNACEAE)

***Vern.:*** Sarpgandha.

***Regional and Other Names :*** *Bengali* – Sarpgandha; *English* - Serpentine root; *Hindi* – Sarpgandha; *Kannada* – Harkaya; *Malayalam* - Amalpori-yan; *Marathi* – Harki; *Sanskrit* – Nakuli, Sarpagandha; *Tamil* - Sovannamil-bori; *Telugu* – Paataalagani.

***Distribution :*** Sub-himalayan tract from Sirhind eastwards, Burma, Andmans, Konkan, W. Ghats of Madras Presidency.

***Taxonomic description :*** A small perennial herbaceous shrub; leaves simple whorled; flowers white-slightly pink in clusters.

***Flowering & Fruiting :*** November-December.

***Medicinal parts :*** Roots, Leaves.

***Taste :*** Bitter.

***Chemical constituents :*** Root contains Arachidic acid, Lauric acid, Oleic acid, Myristic acid, Palmitic acid, Stearic acid, Ajmalcine, Deserpidine, Indobine, Indobinine, Rescinnanidine, Rescinnamine, Rescinnaminol, Reserpentine, Reserpine, Serpagine, Serpentine, Serpentinine, Ajmalimines, Ajmaline, Oleoresin, Serposterol.

***Medicinal properties :***

नाकुली तुवरा तिक्ता कटुकोष्णा विनाशयेत्।
भोगिलूता वृश्चिकाखुविषज्वरकृमिव्रणान् ।। (भा.प्र.)

Nākulī tuwrā ṭiktā katukoṣṇā vināśyeṭ
Bhogilūtā vriścikākhuviṣhjvarkṛimivṛṇāṇ. (Bhā. Pra.)

Root - Analgesic, anti-hypertensive, anti-pyretic, sedative, tonic.

***Medicinal uses :*** Root - Dismennorrhagia, epilepsy, fever, hypertension, insomnia, uterine contraction, wound.

***Ethno- Medicinal uses :*** Root - Abortifacient (Tarefder, 1983); ascites (Sarma *et al.,* 2006); Blood pressure, Insomnia, Nervous disorders (Bhalla *et al.,* 1982); Fever (Nagendra prasad & Abraham, 1984; Das *et al.,* 1983; Maheshwari *et al.,* 1980); Malaria (Kumar *et al.,* 1980; Deka *et al.,* 2006) madness (Narain & Singh, 2006); hypertension (Narain & Singh, 2006; Deokule, 2006); dog bite (Baruah *et al.,* 2006); snake bite, scorpion sting (Kadavul & Jisha, 2006).

Leaf- High blood pressure (Baruah *et al.,* 2006).

## *RICINUS COMMUNIS* L. (EUPHORBIACEAE)

***Vern.:*** Arandi.

***Regional and Other Names :*** *Bengali* – Rehri; *English* - Castor; *Hindi* - Erandi; *Kannada* – Haralu; *Malayalam* – Avanakku; *Marathi* – Erand; Sanskrit Svetaeranda; *Tamil* - Avanakku; *Telugu* – Amudamuchepta.

***Distribution :*** Throughout India.

***Taxonomic description :*** Large shrubs or small trees, with fistular stem and large, palmetly 5 to 10-lobed leaves. Flowers in pyramidal panicles. Capsules globose, spiny, of 3- cocci; seeds caruncled.

***Flowering & Fruiting :*** Throughout the year.

***Medicinal parts :*** Root, Leaves, Flowers, Fruits, Seeds & Seed oil.

***Taste :*** Acrid\ Astringent.

***Chemical constituents :*** Root contains Resin, Tannin, starch.

Seeds contain 50% oil, protein, Ricinine, Ricin, Lipase enzyme, Fixed oil.

Fixed oil contains Glycerides, Ricinolesis, Isoricinoleic acid, stearic acid, Crystals of calcium oxalate, nitrate.

***Medicinal properties :***

श्वेतैरण्ड: सकटुकरसातिक्त उष्ण: कफार्ति ।
ध्वंसं धत्ते ज्वरहरमरूत् कासहारी रसार्ह: ।। (रा. नि.)

Śvetạiraṇdaḥ sakatukarasātikta uṣṇaḥ kaphạ̄rti
Ḍḥvaṇsaṃ dhạtte jvaraharamaruṭ kāsahārī rasạ̄rhaḥ. (Raj. ni.)

Alterative, anti-inflammatory, aphrodisiac, carminative, febrifuge, purgative, stomachic.

***Medicinal uses :*** Fruits - Asthma, bronchitis, epilepsy, eye diseases, febri-fuge, hydrocoel, jaundice, lumbago, piles, rheumatism, sciatica, skin diseases.

Leaves - Cough, worms infestation.

***Ethno- Medicinal uses :*** Leaves - Dyspepsia, gonorrhoea (Bhalla *et al.*, 1982); bronchitis (Lal & Yadav, 1983); piles (Lal & Yadav, 1983; Mishra & Billore, 1983; Shah & Joshi, 1971); skin dieases (Rao, 1981; Nautiyal, 1981); cuts, ear complaint (Shah & Joshi, 1971); pimples, cuts, wounds, boils (Purkyastha & Subhan, 2006); headache (Deokule, 2006; Das *et al.*, 2006); muscular pain (Sarma *et al.*, 2006); bodyache (Das *et al.*, 2006).

Fruits - Laxative (Rajwar, 1983).

Seeds - Labour (Tarafder, 1984); intestinal warm (Jadeja *et al.*, 2006); Gout (Jadeja *et al.*, 2006).

Seed oil – Cough, cold, asthma, cardiac pain (Kumar *et al.*, 2006).

## *RUBIA CORDIFOLIA* L. (RUBIACEAE)

***Syn.:*** *R. munjista* Roxb.

***Vern.:*** Manjit.

***Regional and Other Names :*** *Bengali* – Manjistha; *English* - Indian madder; *Hindi* – Manjit; *Kannada* – Slomalate; *Malayalam* – Manjetti; *Marathi* – Manjeshta; *Sanskrit* –Manjistha; *Tamil* - Manjitti, Shevelli; *Telugu* - Manjistateega, Tamravalli.

***Distribution :*** Throughout India.

***Taxonomic description :*** Perennial climbers; stem acutely 4-angled with minute, recurved prickles on the angles; leaves in whorls of 3-4, ovate or ovate-oblong to lanceolate, glabrous. Flowers greenish-yellow in cymes; fruit glabrous, 1- seeded, bluish black.

***Flowering & Fruiting :*** July – December.

***Medicinal parts :*** Root, stem.

***Taste :*** Bitter, astringent.

***Chemical constituents :*** Root contains Purpurin, manjistin, garancin, purpuroxanthin, resin, glucose, sucrose, triterpenes, lucidin, anthroquinones, fatty acids, gum.

***Medicinal properties :***

मुंजिष्ठा मधुरा तिक्ता कषाया स्वरवर्णकृत् ।
गुरुरूष्णा विषश्लेष्मशोथयोन्यक्षिकर्णरूक् ।।
रक्तातिसार कुष्ठास्त्रवीसर्पव्रणमेहनुत् । (भा. प्र.)

Maṇjiṣthā madhurā tiktā kasāyā ṣvaravaṛṇakṛt
Gururuṣṇā viṣaṣleṣmaśothayoṇyakṣikaṛṇaruk
Raktātisāra kuṣthā ṣṭravīsaṛpavṛaṇa mehanuṭ. (Bhā. Pra.)

Root - Acrid, alexiteric, analgesic, anthelmintic, anti -dysentery, antipyretic, bitter, sweet.

Leaf - Appetiser, sweet.

***Medicinal uses :*** Root - Diseases of the Uterus, the vagina, the eye, the ear, the blood; erysipelas, inflammation, jaundice, leucoderma, piles, ulcers, urinary discharges.

Leaf - Biliousness.

***Ethno- Medicinal uses :*** Leaf - Ulcer (Rao, 1981).

Stem - Snake & scorpion bite (Mukherjee *et al.*, 1986).

Bark- Dysentery, piles (Jadeja *et al.,* 2006).

Root - Chest complaint, Jaundice (Mukherjee *et al.*, 1986); liver complaint (Sharma *et al.*, 1979); menorrhoea (Uniyal, 1968); paralysis (Banerjee, 1977); snake bite (Udayan *et al.,* 2006); headache (Gupta, 2006).

## *SANTALUM ALBUM* L. (SANTALACEAE)

***Vern.:*** Chandan.

***Regional and Other Names :*** *Bengali* – Chandan; *English* - Sandal wood; *Hindi* - Safed chandan; *Kannada* – Srigandha; *Malayalam* – Chandanam; *Marathi* – Chandan; *Sanskrit* - Chandanam; *Tamil* - Sandanam, ulocidam; *Telugu* – Srigandhamu.

***Distribution :*** W. Peninsula. Cultivated elsewhere.

***Taxonomic description :*** Small, evergreen trees, with drooping branches. Leaves ovate-lanceolate. Flowers reddish-brown, in axillary and terminal cymes. Fruits globose, dark purple.

***Flowering & Fruiting :*** September – December.

***Medicinal parts :*** Heart wood, oil.

***Taste :*** Bitter.

***Chemical constituents :*** Heart wood contains Resin, Tanic acid & Volatile oil which is thick, yellow & aromatic. It contains œ & β – santalol, Aldehyde, ketone, santenone, santalone.

***Medicinal properties :***

चन्दनं शीतलं रूक्षं तिक्तमल्हादनं लघु ।
श्रमशोषविषश्लेष्मतृष्णापित्तास्त्रदाहनुत् ।। (भा.प्र.)

Candanaṃ śītlaṃ rukshaṃ tiktmạlhādanaṃ laghu
Sṛamśoṣviṣaśleṣmaṭriṣṇāpịttāṣṭradāhnụt. (Bhā. Pra.)

Wood - Alexeteric, antipyretic, aphrodisiac, bitter, cooling.

***Medicinal uses :*** Wood - Biliousness, bronchitis, burning sensation, heart diseases, small pox, vaginal discharges.

***Ethno- Medicinal uses :*** Seed - Abortifacient (Bhalla *et al.,* 1982).

Wood - Fever, skin diseases (Bhalla *et al.,* 1982); headache (Nagendra Prasad & Abraham, 1984); boils (Lal & Yadav, 1983).

Wood oil- Gonorrhoea, cough, urinary troubles (Deokule, 2006).

Stem bark- Acidity (Jadeja *et al.,* 2006).

Stem- diabetes (Ghosh, 2006); Gout (Jadeja *et al.,* 2006).

## *SARACA ASOCA* (Roxb.) De Wilde. (CAESALPINIACEAE)

***Syn.:*** *S. indica* Auct. non L.

***Vern.:*** Ashok.

***Regional and Other Names :*** *Bengali* – Asoka; *English* - Ashoka tree; *Gujarati* - Ashank; *Hindi* - Asok; *Kannada* – Asokadamra; *Malayalam* – Asokam; *Marathi* – Sitechaashoke; *Sanskrit* – Asoka, Hempusp; *Tamil* - Asogam; *Telugu* – Asokamu.

***Distribution :*** Central and E. Himalaya, E. Bengal, Burma, W. Peninsula.

***Taxonomic description :*** A small evergreen tree. Leaves glabrous cariaceous; leaflets oblong or oblong-lanceolate, slightly oblique. Flowers fragrant. Calyx petaloid, tubular. Anthers purple. Ovary pubescent. Pods black, linear-oblong, veined.

***Flowering & Fruiting :*** January-April.

***Medicinal parts :*** Bark, flower, seed.

***Taste :*** Bitter.

***Chemical constituents :*** Bark contains Glycosides, Saponins, Tannins, Essential oil, Haematoxyline, Ketosterol, Kaempferol, Calcium, Iron, Fatty acids, Gallic acid.

***Medicinal properties :***

अशोकः शीतलस्तिक्तो ग्राही वर्ण्यः कषायकः ।
दोषापची तृषा दाहकृमि शोष विषास्त्रजित् ।। (भा. प्र.)

Asokaḥ śītalaṣtikto grāhī vaṛṇyaḥ kaṣāyakaḥ
Doṣāpacī ṭriṣā dāhakṛimi śoṣa viṣāṣṭrajiṭ. (Bhā. Pra.)

Anthelmintic, anti-inflammatory, astringent, blood purifier, cooling, demulcent, febrifuge, refrigerant, spasmodic, stimulant.

***Medicinal uses :*** Anti-bacterial, anti-cancer, anti-fungal, anti-tumour, cardiac stimulant, disuria, leucoderma, menorrhagia, piles, ulcers.

***Ethno- Medicinal uses :*** Leaf - Heart diseases, menstrual complaint, menorrhoea, urinogenital diseases (Kanick *et al.,* 1981).

Bark - menorrhoea, urinogenital diseases (Shah *et al.,* 1981); dysentery, piles (Jadeja *et al.,* 2006); Menstrual complaint (Shah *et al.,* 1981; Kadavul & Jisha, 2006).

## ***SENNA ABSUS*** (L.) Roxb. (CAESALPINIACEAE)

***Syn.:*** *Cassia absus* L.

***Vern.:*** Chaksusya.

***Regional and Other Names :*** *Gujarati* - Chimar; *Hindi* - Chaksu; *Marathi* – Kankuti; *Sanskrit* –Chakshushya, Kulatthika; *Tamil* - Kattukol; *Telugu* – Chanupalavittulu.

***Distribution :*** Throughout India.

***Taxonomic description :*** Annual herbs, clothed with glandular hairs; rachis grooved; leaflets obliquely elliptic to obovate; flowers pink or red, in terminal or leaf opposed, few-flowered racemes; pods hairy, flat.

***Flowering & Fruiting :*** August – October.

***Medicinal parts :*** Leaves, seeds.

***Taste :*** Bitter (Leaves); Bitter (Seeds).

***Chemical constituents :*** Seed contains Alkaloids, sugars, fats, cannins, mucilage, major protein, mangenese.

***Medicinal properties :***

कुलत्थिका हिमा श्लेष्मविषनेत्रामयापहा ।
शाकं वन्यकुलत्थस्य हिक्काभिष्यन्दनाशनम् ।। (कै. नि.)

Kulaṭthika hima sleṣmavisaneṭramayapaha
Sakaṃ vaṇyakulaṭthaṣya hikkabhiṣyaṇdanasanaṃ. (Kai. Ni.)

Leaf - Acrid, astringent, bitter, constipating, expectorant, haematinic.

Seed - Acrid, astringent, bitter, constipating, diuretic, haematinic, vulnerary.

***Medicinal uses :*** Leaf - Anaemia, asthma, cough, diarrhoea, hiccough, nasitis, tumours.

Seed - Diarrhoea, dysentery, haemorrhages, obesity, poisonous bites, ulcers, wounds.

***Ethno- Medicinal uses :*** Leaf - Cough (Sharma *et al.,* 1979); wound (Shah *et al.*, 1981).

Seed - Eye infection (Bhatt & Sabnis, 1987); skin diseases (Sharma *et al.,* 1979).

## *SENNA AURICULATA* (L.) Roxb. (CAESALPINIACEAE)

***Syn.:*** *Cassia auriculata* Roxb.

***Vern.:*** Tarwar.

***Regional and Other Names :*** *English* - Tanner's cassia; *Gujarati* - Aval; *Hindi* - Tarvar; *Malayalam* – Aviram; *Marathi* – Taravada; *Sanskrit* - Avartaki; *Tamil* - Avarai; *Telugu* –Tangera

***Distribution :*** Dry regions of the Country.

***Taxonomic description :*** Finely pubescent shrubs, leaflets 5 – 12 pairs, oblong-obovate, obtuse or emarginate, apiculate; flowers in corymbose racemes, yellow; pods oblong, mucronate, deeply impressed and septate between the seeds.

***Flowering & Fruiting :*** August– November.

***Medicinal parts :*** Root, bark, leaves, flowers, seeds.

***Taste :*** Bitter.

***Chemical constituents :*** Plant contains Tannins.

White bark contains More tannic acid.

***Medicinal properties :***

आवर्त्तकी कषायाम्ला शीतला पित्तहारिणी ।। (रा. नि.)

Āvarṭtakī kasāyāṃlā śītalā piṭtahāriṇī. (Ra. Ni.)

Root - Astringent, cooling, depurative.

Leaf - Anthelmintic, depurative.

Seed - Anthelmintic, aphrodisiac, astringent, constipating, cooling, depurative, sour, stomachic.

***Medicinal uses :*** Root - Asthma, leprosy, skin disease, tumours.

Leaf - Leprosy, skin diseases, ulcers.

Flower - Diabetes.

Seed - Abdominal disorders, diabetes, diarrhoea, dysentery, leprosy, ophthalmia, skin diseases, swelling.

***Ethno- Medicinal uses :*** Root - Rheumatism (Sharma & Malhotra, 1984).

Root bark- Constipation, dysentery (Jadeja *et al.,* 2006); Gout (Jadeja *et al.,* 2006).

Stem - Toothache (Sharma & Malhotra, 1984).

Leaf- Swelling (Deokule, 2006).

Seed - Cough & asthma (Singh & Panedy, 1980); skin diseases (Baruah *et al.,* 1987).

Whole plant - Dysentery (Malhotra & Moorthy, 1973).

## *SENNA OCCIDENTALIS* (L.) Link. (CAESALPINIACEAE)

***Syn.:*** *Cassia occidentalis* L.

***Vern.:*** Kasmard.

***Regional and Other Names :*** *Bengali* - Kalkashunda; *English* - Negro coffee; *Gujarati* - Kasodari; *Hindi* - Kasaumdi; *Kannada* - Doddaagace; *Malayalam* - Ponnariviram; *Marathi* - Kasoda; *Sanskrit* - Kasamardah; *Tamil* - Nattam takarai; *Telugu* – Kasinda.

***Distribution :*** Throughout India.

***Taxonomic description :*** Diffuse undershrubs, with 3-6 pairs of leaflets and a gland on the petiole. Flowers yellow, in corymbose panicles. Pods linear-oblong.

***Flowering & Fruiting :*** October – March.

***Medicinal parts :*** Whole plant.

***Taste :*** Bitter.

***Chemical constituents :*** Plant contains Emodin, Tannic acid, fatty oils, Mucilage, aminoacids, Chrysophanol, Sitosterol, Xanthon, Cassiodin, Nitrogen bases, Betain, Choline, Stachydrine, and Trigonelline.

Seed contains Galactomanon.

***Medicinal properties :***

कासमर्ददलं रुच्यं वृष्यं कासविषास्त्रनुत् ।
मधुरं कफवातघ्नं पाचनं कण्ठशोधनम् ।।
विशेषतः कासहरं पित्तघ्नं ग्राहकं लघु । (भा.प्र.)

Kāsamardadalaṃ rucyaṃ vṛṣyaṃ kāsvishāṣṭranuṭ
Madhuraṃ kaphavātaghnaṃ pācanaṃ kaṇthaśodhanaṃ
Viśesataḥ kāsaharaṃ piṭtaghnaṃ grāhakaṃ laghu. (Bha.Pra.)

Root - Acrid, anti – inflammatory, bitter, digestive, diuretic, stomachic.

Leaf - Acrid, aphrodisiac, bitter, depurative, expectorant.

Seed - Acrid, bitter, depurative, diuretic, expectorant, febrifuge, purgative, stomachic.

***Medicinal uses :*** Root - Colic, diabetes, dyspepsia, elephantiasis, epilepsy, flatulence, inflammation, ringworm, scorpion sting.

Leaf - Asthma, bronchitis, cough, fever, hiccough, hydrophobia, leprosy, pruritus, ulcers, wounds.

Seed - Bronchitis, constipation, cough, dyspepsia, erysipelas, fever, flatulence, hiccough, leprosy, stranguray, ulcers.

***Ethno- Medicinal uses :*** Leaf - Bone fracture (Apparanantham & Chelladurai, 1986); eczema, ringworm, throat infection (Shah *et al.*, 1983); fever (Tarafder, 1986); skin diseases (Malhotra & Moorthy, 1973; Das *et al.*, 2006); wound (Singh *et al.*, 1987); stomach trouble (Ganeshan *et al.*, 2006); rheumatism (Kumar *et al.*, 2006).

Root - Gastric complaint (Goel *et al.*, 1984); bodyache, urinary problems (Narain & Singh, 2006b); skin diseases (Jain *et al.*, 1973); whooping cough (Singh & Pandey, 1980).

Seed - Skin diseases (Maheshwari & Singh, 1987; Das *et al.*, 2006); whooping cough (Singh & Paney, 1980).

## *SENNA TORA* (L.) Roxb. (CAESALPINIACEAE)

***Syn.:*** *C. tora* L.; *C. obtusifolia* L.; *C. toroides* Roxb.; *C. foetida* Pers.

***Vern.:*** Chakramard.

***Regional and Other Names :*** *Bengali* – Chakunda; *English* - Foetid cassia, wild senna; *Gujarati* - Kawario; *Hindi* - Cakunda, Punwar; *Kannada* - Taragasi, Gandutogache; *Malayalam* -Takara, Chakramandarakam; *Marathi* - Takala, tarota; *Sanskrit* - Cakramardah; *Tamil* - Tagarai; *Telugu* – Tantiyamu.

***Distribution :*** Throughout India.

***Taxonomic description :*** Suffruticose herbs. With 3 pairs of leaflets and glands between them. Flowers yellow or reddish tinged, geminate; all 7 anthers rounded at apex. Pods linear, turgid.

***Flowering & Fruiting :*** September- November.

***Medicinal parts :*** Leaves, seeds.

***Taste :*** Bitter.

***Chemical constituents :*** Seed contains Glucosides, (Emodin) proteins, steroids, Amino acids, Reducing sugars, Triterpenes, Tannins, oleic acid, Linotic acid, Palmitic acid, Lignocenic acid, Rubrofusarin, Fixed oils, sodium, Potassium & Iron.

***Medicinal properties :***

चक्रमर्दः प्रपुन्नाटो दद्रुघ्नो मेषलोचनः ।
पद्माटः स्यादेत्गजश्चक्री पुन्नाट इत्यपि ।। (भा.प्र.)

Cakṛamaṛdaḥ pṛapuṇnāto dadṛughno mesalocanaḥ
Padmātaḥ ṣyādeṭgajaścakṛi puṇnāta iṭyapi. (Bha. Pra.)

Leaves & seeds - Acrid, anthelmintic, antiperiodic, depurative, expectorant, laxative, ophthalmic.

***Medicinal uses :*** Leaves & seeds - Bronchitis, cardiac disorders, colic, constipation, cough, dyspepsia, flatulence, intermittent fever, leprosy, pruritus, ringworm, skin diseases.

***Ethno- Medicinal uses :*** Leaf - Stomach trouble; Indigestion (Tarafder, 1983); Nightblindness (Singh & Maheshwari, 1983); Bone fracture (Jain *et al.*, 1973); Boils (Sexena & Vyas, 1981); Jaundice (Hemadri & Rao, 1984); Skin diseases (Shah *et al.*, 1981; Purkayastha & Subhan, 2006; Kumar *et al.*,

2006; Yesodharan & Sujana, 2006; Das *et al.,* 2006); Wound (Jain, 1963; Yesodharan & Sujana, 2006); dysentery (Ganeshan *et al.*, 2006); worms (Kadavul & Jisha, 2006); cuts (Yesodharan & Sujana, 2006).

Seeds - Eczema, Ringworm (Shah *et al.,* 1983); Cold (Shah & Joshi, 1971); Itching (Sexena & Vyas, 1981); Scorpion bite (Nagar, 1985); Stomach trouble (Rajwar, 1983); snake bite (Kiruba *et al.*, 2006); skin diseases (Kumar *et al.,* 2006; Das *et al.,* 2006); worms (Kadavul & Jisha, 2006); Gout (Jadeja *et al.,* 2006).

Root - Ringworm (Bhalla *et al.,* 1982); irregular menstruation (Sarma *et al.,* 2006); Skin diseases (Janardhanan, 1963); Fever (Tarafder, 1987); Scorpion bite (Maheshwari *et al.,* 1986); snake bite (Kumar *et al.,* 2006).

Fruit- Anthelmintic (Deokule, 2006).

## ***SESBANIA GRANDIFLORA*** (L.) Poir. (PAPILIONACEAE)

***Syn.:*** *Agati grandiflora* Desv.; *Robinia grandiflora* L.

***Vern.:*** Agastya.

***Regional and Other Names :*** *Bengali* – Agati; *English* – Sesban; *Gujarati* – Agasthio; *Hindi* – Agasti; *Kannada* – Agase; *Malayalam* – Akatthi; *Marathi* – Agasta; *Sanskrit* – Agasti; *Tamil* – Agathi; *Telugu* – Avesi.

***Distribution :*** Indigenous from Malaya to N. Australia; cultivated in many parts of India.

***Taxonomic description :*** Tall slender tree; Leaves pinnately compound, leaflets 16-30 pairs, linear-oblong. Flowers white.

***Flowering & Fruiting :*** December-march.

***Medicinal parts :*** Root, Bark, Leaves, Flowers, Fruits.

***Taste : -*** Bitter

***Medicinal properties :***

अगस्तिपुष्प चुर्णेन महिषं जनयेद्दधि ।
तदुत्थनवनीतेन देहजं स्फुटनं जयेत् ।। (भा. प्र.)

Agaṣtpuṣpa chuṛṇane mahiṣaṃ janyeḍdhi
Taduṭhanavnītane dehajaṃ ṣphutanaṃ jayeṭ. (Bhā. Pra.)

Bark - Astringent.

Leaves - Alexeteric, anthelmintic, bitter, indigestible.

Flowers - Appetizer, cooling.

Fruits - Acrid, alexeteric, bitter, laxative.

***Medicinal uses :*** Leaves - Epilepsy, gout, itching, leprosy, night blindness.

Flowers - Biliousness, bronchitis, gout, night blindness.

Fruits - Anaemia, bronchitis, fevers, pains, tumours.

***Ethno- Medicinal uses :*** Leaf - Ulcer (Ganesan *et al.,* 2006); Mumps (Sudhakar & Rolla, 1985); headache (Sudhakar & Rolla, 1985; Deokule, 2006); eye diseases (Rajendran & Rengamani, 2006).

## *SIDA CORDIFOLIA* L. (MALVACEAE)

***Vern.:*** Bala.

***Regional and Other Names :*** *Bengali* - Swetberela, kureta; *English* - Country mallow; *Gujarati* - Baladana; *Hindi* - Bariara, kharenta; *Kannada* - Hettuthi, hettugigada; *Malayalam* - Kurunthotti, velliurum; *Marathi* – Chikana; *Sanskrit* - Bala; *Tamil* - Karumthotti, milatutti; *Telugu* - Chirwbenda, tellantisa.

***Distribution :*** Tropical and sub-tropical regions of both hemispheres.

***Taxonomic description :*** Undershrubs, with ovate-cordate leaves and axillary yellow flowers. Mericarps 10, 2-awned.

***Flowering & Fruiting :*** August– November.

***Medicinal parts :*** Root, leaves, seeds, whole plant.

***Taste :*** Bitter.

***Chemical constituents :*** Root contains Beta-sitosterol, alkaloids-visicinonoe, vasicine, vasicinol, choline, & betain.

Seeds contain Ephedrine, fatty oils, steroids, resin, mucin, sterculic acid, malvalic acid, linoleic acid, potassium nitrate.

***Medicinal properties :***

बलाचतुष्टयं शीतं मधुरं बलकान्तिकृत् ।
स्निग्धं ग्राहि समीरास्त्रपित्तास्त्रक्षतनाशनम् ।। (भा.प्र.)

Balācatuṣtyaṃ śītaṃ madhuraṃ balkāṇtikṛuṭ
Ṣnigdhaṃ grāhi samīrāṣṭrapiṭtāṣṭrakṣatnāśnaṃ. (Bhā. Pra.)

Plant - Aphrodisiac, astringent, bitter, emollicent, sweet, tonic.

Fruits - Acrid, aphrodisiac, astringent, cooling, digestive, sweet.

***Medicinal uses :*** Plant - Cough.

Bark - Urinary trouble & discharges.

Fruits - Blood diseases, diseases of throat, piles.

***Ethno- Medicinal uses :*** Root - Boils (Jain *et al.,* 1973); wound (Das *et al.,* 1983; Kumar *et al.,* 2006); stomatitis (Shah & Gopal, 1986); leucorrhoea (Upadhyay & Jaswant Singh, 2005); post delivery trouble (Sarma *et al.,* 2006); colic pain (Chatterjee & Bhatterjee, 2006).

Whole plant– Dysentery (Singh & Mahshwari, 1983); rheumatism (Malhotra & Moorthy, 1973); fever, leucorrhoea, nervous disorders (Kumar *et al.,* 2006); gonorrhoea (Singh & Mahshwari, 1983; Kumar *et al.,* 2006); spermatorrhoea (Prakash & Singh, 2006.).

Seed - Gonorrhoea, leucorrhoea (Singh, 1945).

Leaf - Gonorrhoea, sex weaknes (Tarafder & Chaudhari, 1981); wounds & cuts (Kiruba, 2006).

## *SIDA RHOMBIFOLIA* L. (MALVACEAE)

***Vern.:*** Bala.

***Regional and Other Names :*** *Bengali* - Lal–berela; *English* - Common bale; *Gujarati* – Baladana; *Hindi* - Bariara, lal berela; *Kannada* - Kallangadale, bolamgadale; *Malayalam* - Anakurunthotti, vatturum; *Marathi* – Sahadevi; *Sanskrit* –Mahabala; *Tamil* - Sirramutti, chitra mutti; *Telugu* - Athibala, chettu.

***Distribution :*** Tropical and sub-tropical regions of both hemispheres.

***Taxonomic description :*** Much-branched herbs, with obovate- rounded or rhomboid leaves and axillary, solitary, yellow flowers. Mericarps 1-awned or awnless.

***Flowering & Fruiting :*** October – December.

***Medicinal parts :*** Whole plant, root, leaves.

***Taste :*** Bitter.

***Chemical constituents :*** Plant contains Ephedrine, cellulose, lignin, quanozolines, β- Phenethylamines, Tryptomine, Sterculic acid, Malvalic acid, Linoleic acid.

***Medicinal properties :***

महाबलामूलहौषधाभ्यां क्वाथो निहन्याद्विषमज्वरं हि ।
शीतं सकम्पं परिदाहयुक्तं विनाशयेत् द्वित्रिदिनप्रयोगात् ।। (भा.प्र.)

Mahābalāmūlhōṣdhābhyāṃ kwātho nihaṇyāḍviṣmjvaraṃ hi
Śītaṃ sakaṃpaṃ paridāhyuktaṃ vināśyeṭ dviṭridinpṛayogāṭ.
(Bhā. Pra.)

Root - Aphrodisiac, sweet, tonic.

Leaf - Aphrodisiac, sweet, tonic.

***Medicinal uses :*** Root - Burning sensation, fever, heart diseases, inflammation, Piles, strangury, urinary complaints.

Leaf - Burning sensation, fever, heart diseases, inflammation, Piles, strangury, urinary complaints.

***Ethno- Medicinal uses :*** Whole plant - Gonorrhoea (Tarafder, 1984); rheumatism, urinary diseases (Das *et al.,* 2006).

Leaf - Dysentery (Sebastian & Bhandari, 1984); Swelling (Malhotra & Moorthy, 1973); wound (Upadhye *et al.,* 1986); spermatorrhoea (Prakash & Singh, 2006.).

Root - Rheumatism (Nagendra Prasad & Abraham, 1984); boils (Tarafder, 1984); asthma & fever (Singh & Pandey, 1980); leucorrhoea (Upadhyay & Jaswant Singh, 2005).

## ***SOLANUM INDICUM*** L. (SOLANACEAE)

***Syn.:*** *S. violaceum* Ortega.; *S. sodomenum* L. non ambig.; *S. anguivi* Lam.

***Vern.:*** Bruhati.

***Regional and Other Names :*** *Bengali* – Bagaun; *English* - Poison berry; *Gujerati* - Mhotiringni; *Hindi* - Barhanta; *Malayalam* – Cheruchunta; *Marathi* – Dorli; *Sanskrit* - Brhati; *Tamil* - Kandal, mundagam; *Telugu* - Chittimulaga, kakimachi.

***Distribution :*** Throughout India.

***Taxonomic description :*** Erect, stout, stellately hairy, armed perennial undershrub; leaves ovate-oblong, subacute or obtuse apex, stellately pubescent. Flowers purple in lateral, cymose corymbose; berries globose, orange yellow.

***Flowering & Fruiting :*** July – December.

***Medicinal parts :*** Fruits.

***Taste :*** Bitter.

***Chemical constituents :*** Fruits contain Glycoalkaloids (Solasonine), Solanine, Diosgenin, Carbohydrates, Maltose, saccharase, melibase, solamargine, solasodine, tometidenol, fatty acids.

Leaves contain Solasonine, diosgenin, β sitosterol, solamargine, solasodine.

***Medicinal properties :***

वृहती ग्राहिणी हृद्या पाचनी कफवातजित् ।
कटुतिक्ताऽऽस्यवैरस्यमलारोचकनाशिनी ।
उष्णा कुष्ठज्वरश्वासशूलकासाग्निमान्धजित् ।। (भा.प्र.)

Bṛhatī grāhiṇī hṛdyā pācinī kaphavātajit
Katutiktāaṣyavaịraṣyamalārocaknāśinī
Uṣṇā kuṣthajvarśvāsaśūlakāsāgnimāṇdyajit. (Bhā. Pra.)

Root - Anthelmintic, astringent, bitter, digestive, pungent.

Fruits - Anthelmintic, bitter, pungent.

***Medicinal uses :*** Root - Leucoderma, fever, asthma, bronchitis, vomiting, pruritus.

Fruits - Pruritus, leucoderma, bronchitis, fever, vomiting, eye diseases.

***Ethno- Medicinal uses :*** Leaf - Stomatitis (Karnick *et al.,* 1981).

Root - Lung diseases (Karnick *et al.,* 1981); asthma, fever, worms (Sudhakar & Rolla, 1985); cough (Malhotra & Moorthy, 1973).

## *SOLANUM NIGRUM* L. (SOLANACEAE)

***Syn.:*** *S. subrum* Mill. Ex Wight.; *S. incertum* Dunal ex Grahum.

***Vern.:*** Makoi.

***Regional and Other Names :*** *Bengali* - Kakmachi, Gurukamai; *English* - Black night shade; *Gujarati* - Piludi; *Hindi* - Gurkamai, Markol; *Kannada* – Ganika; *Malayalam* – Mulagthakkali; *Marathi* – Kangani; *Sanskrit*-Kakamaci; *Tamil*-Munathakkali; *Telugu* - Kachi, kachchipundu.

***Distribution :*** Throughout India.

***Taxonomic description :*** Unarmed, glabrous or pubescent herbs, with ovate-oblong leaves, white flowers and globose, yellow, red or purple fruits.

***Flowering & Fruiting :*** June – February.

***Medicinal parts :*** Whole plant, flowers, leaves, fruits.

***Taste :*** Bitter, acrid.

***Chemical constituents :*** Plant contains Alkaloids, Reducing sugars, glycosides, saponins, steroids, leutein, Lycopene, Vitamin – C, glucose, fructose, caffeic-solasodine, tomatidenel, solamargine, solasomine, trigogenine, potassium, Iron, sulphur, Calcium, Phosphorus.

***Medicinal properties :***

काकमाची त्रिदोषघ्नी स्निग्धोष्णा स्वरशुक्रदा ।
तिक्ता रसायनी शोथकुष्ठार्शोज्वरमेहजित् ।। (भा.प्र.)

Kākmācī ṭridoṣaghnī ṣnigdhoṣṇā ṣvarśukṛadā
Tiktā rasāyanī śothakuṣthāṛśojvarmehjịt. (Bhā. Pra.)

Fruits - Alterative, aphrodisiac, appetiser, bitter, diuretic, laxative, pungent.

***Medicinal uses :*** Fruits - Asthma, bronchitis, cough, dysentery, erysipelas, eye diseases, fever, heart diseases, hiccough, inflammation, itching, leucoderma, piles, urinary discharges, vomiting, worms in the ear.

***Ethno- Medicinal uses :*** Seed - Gouitre (Dar *et al.,* 1984); liver trouble (Das *et al.,* 2006).

Whole plant - Fever (Shekhawat & Anand, 1984); jaundice (Billore, & Audichya, 1978; Dutta Choudhury & Shil, 2006); piles, swelling (Deokule, 2006); diarrhoea (Kataria & Gauchan, 2006); dysentery (Deokule, 2006; Kataria & Gauchan, 2006).

Fruits - Diarrhoea, fever (Bhalla *et al.,* 1982); cough & throat trouble (Megoneitso & Rao, 1983); swelling, (Shah *et al.,* 1983); liver trouble (Purkayastha & Subhan, 2006; Baruah *et al.,* 2006); indigestion (Baruah *et al.,* 2006); stomach pain (Arinathan *et al.,* 2006).

Leaf - Swelling (Maheshwari & Singh, 1984; Purkayastha & Subhan, 2006), boils, skin diseases, earache (Sebastian & Bhandari, 1984); liver complaint (Lal & Yadav, 1983; Baruah *et al.,* 2006); rheumatism (Malhotra & Morrthy, 1973); dysentery (Maheshwari & Singh, 1987); stomach trouble (Abraham, 1981); piles (Singh,1988); eye complaint (Shah & Joshi, 1971); indigestion (Baruah *et al.,* 2006); constipation (Das *et al.,* 2006).

## *SOLANUM SURATTENSE* Burm. f. (SOLANACEAE)

***Syn.:*** *S. maccanni* Sant.; *S. xanthocarpum* Schrad. & Wendl.; *S. diffusum* Roxb.; *S. armatum* Br.; *S. virginianum* Jacq.

***Vern.:*** Bhuri rindi.

***Regional and Other Names :*** *Bengali* – Kantakari; *English* - Yellow berried, night shade; *Gujarati* - Bhoyaringani; *Hindi* - Kateli, katal; *Kannada* – Kandan; *Malayalam* – Karychunda; *Marathi* - Kateringanii, bhuiringani; *Sanskrit*-Kantakari; *Tamil*-Kandankattiri; *Telugu* - Pinnamulaka, nelamulakaa.

***Distribution :*** Throughout India.

***Taxonomic description :*** Prostrate, prickly herbs, with sinuately lobed to subpinnatifid leaves. Flowers violet-purple or white. Fruits globose, yellow when ripe.

***Flowering & Fruiting :*** Feburary – June.

***Medicinal parts :*** Whole plant, fruits, seeds.

***Taste :*** Bitter, acrid, pungent.

***Chemical constituents :*** Plant contains Alkaloids (Solasonine), diosgenin. Solacinine.

Seed contains Essential oil, solanocarpine.

Flower contains Flavonol, glycoside, queretin, apigenin, sitosterol.

Leaves contain Coumarine, scopolin, esculin, esculetin.

***Medicinal properties :***

कण्टकारी सरा तिक्ता कटुका दीपनी लघुः ।
रूक्षोष्णा पाचनी कास श्वासज्वरकफानिलान् ।। (भा.प्र.)

Kaṇtakārī sarā tiktā katukā dīpanī laghuḥ
Rukṣoṣṇā pāchanī kāsa śvāsajvara kaphā nilāṇ. (Bhā. Pra.)

Root - Anthelmintic, appetiser, bitter, laxative, pungent, stomachic.

Fruits - Anaphrodisiac, anthelmintic, appetiser, bitter, digestive.

***Medicinal uses :*** Root - Asthma, bronchitis, fever, heart diseases, lumbago, pains, piles, strangury.

Fruits - Asthma, fever, heart diseases, pruritus.

***Ethno- Medicinal uses :*** Seed - Blisters, boils (Maheshwari & Singh, 1987); toothache (Jain & De, 1966).

Leaf - Eye complaint (Saxena & Vyas, 1983).

Flower - Blister, boils (Maheshwari & Singh, 1984).

Whole plant - Asthama, cough, dropsy, gonorrhoea, fever (Bhalla *et al.*, 1982); toothache (Jain & De, 1966); swelling (Virjee *et al.*, 1984).

Root - Asthama, fever (Shah *et al.*, 1983); bronchitis (Mahaswari & Singh, 1984); cold, cough, scorpion bite (Maheshwari *et al.*, 1986); toothache (Gupta, 1981); cough (Upadhyay & Jaswant Singh, 2005).

Fruits - Asthama (Nagendra Prasad & Abraham, 1984); piles (Shekhawat & Anand, 1984); paralysis (Sudhakar & Rolla, 1985); gum trouble, toothache (Shah *et al.*, 1983); earache (Jain, 1963); cough (Jain & De, 1966).

## *SYMPLOCOS RACEMOSA* Roxb. (SYMPLOCACEAE)

***Vern.:*** Lodhra.

***Regional and Other Names :*** *Bengali* – Lodh; *English* - Lodh tree; *Hindi* - Lodh; *Kannada* - Balaloddu-ginamara; *Malayalam* – Pachotti; *Marathi* – Lodh; *Sanskrit*- Lodhra; *Tamil* - Vellilethi; *Telugu* – Lodhugo.

***Distribution :*** Throughout N.-E. India.

***Taxonomic description :*** A small tree about 20 ft. high with greyish bark; Leaves elliptic-ovate, crenate, and glabrous; Flowers aromatic, yellow; Drupe ellipsoid.

***Medicinal parts :*** Stem bark.

***Taste :*** Acrid.

***Chemical constituents :*** Bark contains Alkaloids: Leupeorin, Leupeoridin, Glycosides, Loturin, Loturidin, Colloturine.

***Medicinal properties :***

लोध्रो ग्राही लघुः शीतःचक्षुष्यः कफपित्तनुत् ।
कषायो रक्तपित्तासृग्ज्वरातीसारशोथहृत् ।। (भा.प्र.)

Lodhro grāhī laghuḥ śītaḥcakṣuṣyaḥ kaphapittanuṭ
Kaṣāyo raktapittā sṛgjvarātīsāra śothahṛt. (Bhā. Pra.)

Anti-inflammatory, antipyretic, astringent, blood purifier, cooling, styptic, tonic.

***Medicinal uses :*** Diarrhoea, dysentery, elepantiasis, eye troubles, fever, leprosy, leucorrhoea, liver troubles, piles, ulcer.

***Ethno- Medicinal uses :*** Stem - Miscarriage (Tarafder, 1984).

Root - Blood purifier (Tarafder, 1987).

## *SYZYGIUM AROMATICUM* (L.) Merr. & Perr. (MYRTACEAE)

***Syn.:*** *Eugenia caruophyllata* Thumb.

***Vern.:*** Laung.

***Regional and Other Names :*** *Bengali* – Laung; *English* - Clove; *Gujarati* – Lavang; *Hindi* - Laung; *Kannada* – Laung; *Malayalam* – Krambu; *Marathi* – Laung; *Sanskrit* - Lavang; *Tamil* - Krambu; *Telugu* – Lavangalu.

***Taxonomic description :*** A large shrub or medium sized tree; Leaves simple, opposite; Flowers small, pale-purple.

***Medicinal parts :*** Dried flower bud with peduncle.

***Taste :*** Bitter-Acrid.

***Chemical constituents :*** Dried flower bud contain Beta-caryophyllene, Eugenol, Acetate, Methyl Salicylate, N-Amylcarbinol, Benzyl alcohol, Furfural, Furfury alcohol, Vanillin, Eugenitine, Eugenon, Galactose, Glucose, Fructose, Rhamnose, Sucrose, Gallotanic acid.

***Medicinal properties :***

लवंगं कुसुमं हृद्यं शीतलं पित्तनाशनम् ।
चक्षुष्यं विषहृद् वृष्यं मांगल्यं मूर्धरोगहृत् ।। (ध. नि.)

Lavaṇgaṃ kusumaṃ hṛdyaṃ śitalaṃ piṭtanāśanaṃ
Caksusyaṃ viṣahṛḍ vṛṣyaṃ māṇgaḷyaṃ mūṛdhroghṛt. (Dha. ni.)

Dried flower bud - Appitizer, aromatic, carminative, stimulant, stomachic.

***Medicinal uses :*** Dried flower bud - Anthelmintic, asthma, blood purifier, cold, colic, cough, diuretic, flatulence, tuberculosis.

***Ethno- Medicinal uses :*** Cloves - Anti- fertility (Tarafder, 1983); post delivery cases (Jadhav, 2006).

## ***SYZYGIUM CUMINI*** (L.) Skeels (MYRTACEAE)

***Syn.:*** *Eugenia jambolana* Lam.

***Vern.:*** Jamun.

***Regional and Other Names :*** *Bengali* – Katajam; *English* – Jambu fruits, black plum; *Gujarati* - Jambu; *Hindi* - Jamun; *Kannada* – Nerali; *Malayalam* - Naval, nagapazham; *Marathi* – Jambhool; *Sanskrit* - Jambula; *Tamil* - Navel peru, negapazham; *Telugu* - Jamburu, neredupandu.

***Distribution :*** Throughout India.

***Taxonomic description :*** Trees, with elliptic-lanceolate leaves and pale-green or white flowers in axillary paniculate cymes. Fruits globose or oblong, purple when ripe.

***Flowering & Fruiting :*** March – May.

***Medicinal parts :*** Leaves, bark, fruits, seed.

***Taste :*** Sweet.

***Chemical constituents :*** Bark contains Friedelin, kaempferol, tannins, quercetin, β- sitosterol, betullinic acid.

Fruits contains Anthocyanin, eugin.

Seed contains Ellagic acid, IAA, Oxalic acid, citric acid, glycollic acid, glucose, fructose, gallic acid.

Leaves contain Glycine, alanin, leucin, tyrosin.

***Medicinal properties :***

राजजम्बुफलं स्वादु विषटिम्भि गुरु रोचनम् ।। (भा.प्र.)

Rājjaṃbuphalaṃ ṣwādu viṣtiṃbhi guru rocnaṃ. (Bhā. Pra.)

Bark - Acrid, anthelmintic, digestive, sweet.

Fruits - Acrid, astringent, cooling, sweet.

Seed - Astringent, sweet.

***Medicinal uses :*** Bark - Asthma, biliousness, blood diseases, bronchitis, dysentery, Sore throat, ulcer.

Fruits - Biliousness.

Seed - Diabetes.

***Ethno- Medicinal uses :*** Seed - Diabetes (Shah *et al.,* 1981; Deokule, 2006; Rajendran & Rengamani, 2006; Das *et al.,* 2006); Pimples (Roy & Chaturvedi, 1986).

Leaf - Dysentery (Lal & Yadav, 1983); diabetes (Purkyastha & Subhan, 2006; Das *et al.,* 2006); wound, itching (Sarma *et al.,* 2006).

Bark - Dysentery (Joshi, 1982; Deokule, 2006; Kataria & Gauchan, 2006; Sharma *et al.,* 2006); Piles (Singh, 1988); diabetes (Purkyastha & Subhan, 2006); diarrhoea (Deokule, 2006; Kataria & Gauchan, 2006).

Fruits - Stomatitis (Mahshwari & Singh, 1984).

Root- Dysentery (Jadeja *et al.,* 2006).

## *TAMARINDUS INDICA* L. (CAESALPINIACEAE)

***Syn.:*** *T. officinalis* Hook.; *T. occidentalis* Gaertn.

***Vern.:*** Imli.

***Regional and Other Names :*** *Bengali* – Tentul; *English* - Tamarind; *Gujarati* - Ambli; *Hindi* - Imli; *Kannada* - Hunise chigun, huli, amli; *Malayalam* – Amlam; *Marathi* – Chinch; *Sanskrit* - Amlika; *Tamil* - Puli, amilan; *Telugu* - Chinta chettu.

***Distribution :*** Throughout India.

***Taxonomic description :*** Evergreen trees, with 6-12 pairs of leaflets and yellow with purple streaks. Pods subtorulose, falcate.

***Flowering & Fruiting :*** April – June.

***Medicinal parts :*** Leaves, stem bark, fruits, seeds, flowers.

***Taste :*** Acidic.

***Chemical constituents :*** Polysaccharides, balsamine, catechin, nasturtium, tamarin, phosphatidic acid and its choline, ethanollamine, serine, inositol, alkaloid, citric acid, tartaric acid, potassium bitartrate.

***Medicinal properties :***

अम्लिकाऽम्ला गुरुवातहरी पित्तकास्त्रकृत् ।
पक्वा तु दीपनी रूक्षा सरोष्णा कफवातनुत् ।। (भा.प्र.)

Aṃlikāaṃlā gururvātharī pịttkāṣṭrakṛuṭ.
Pakvā tu dīpanī ruksā saroṣṇā kaphavātanuṭ. (Bhā. Pra.)

Bark - Astringent.

Leaf - Bitter, laxative, sweet.

Flower - Acrid, appetiser, sour, sweet.

Fruits - Anthelmintic, appetiser, sour, sweet.

***Medicinal uses :*** Bark - Gonorrhoea, paralysis, urinary discharges.

Leaf - Blood diseases, earaehe, eye diseases, inflammation, ophthamia, ringworm, small pox, snake bite, swelling, tumor.

Flower - Urinary discharges.

Fruits - Wound fractures.

Seed - Ulcers, vaginal discharges.

***Ethno- Medicinal uses :*** Fruits - Sun stroke (Shekhawat & Anand, 1984); carminative (Deokule, 2006); swelling (Rajendran & Rengamani, 2006).

Leaf - Fever, urinary complaint (Bhalla *et al.*, 1982); stomatitis (Tarafder, 1983); swelling (Shah *et al.*, 1983); eye complaint (Maheshwari *et al.*, 1986); toothache (Shah & Gopal, 1986); gum problem (Aditya *et al.*, 2006); vomiting (Hembrom & Goel, 2005); rheumatism (Das *et al.*, 2006).

Bark - Stomatitis (Tarafder, 1983).

Seed - Scorpion sting (Ganeshan *et al.*, 2006; Das *et al.*, 2006); snake bite (Das *et al.*, 2006).

## *TECTONA GRANDIS* L. f. (VERBENACEAE)

***Vern.:*** Sagwan.

***Regional and Other Names :*** *Bengali* – Segun; *English* - Teak; *Gujarati* - Saga; *Hindi* - Sagun; *Kannada* - Sagwani, jadi; *Malayalam* – Thekku; *Marathi* – Sag; *Sanskrit* – Anila; *Tamil* - Tekku; *Telugu* – Teku.

***Distribution :*** Konkan, W. Ghats of Bombay and Madras Presidencies, Central India.

***Taxonomic description :*** Deciduous trees, with broadly elliptic large leaves. Flowers white or pinkish, stamens all equal. Fruits enclosed in bladder-like inflated calyx.

***Flowering & Fruiting :*** September–December.

***Medicinal parts :*** Whole plant, bark, wood, leaves, flowers, seeds.

***Taste :*** Bitter.

***Chemical constituents :*** Tactograndon, Betulin, Lupeol, Betulinic acid, Betulinaldehyde, Carbohydrates, fatty acids, lapachanone, deoxylapachol, tectoquinine.

***Medicinal properties :***

शाकः कषायाशिशिरो रक्तपित्तप्रसादनः ।
कुष्ठश्लेष्मानिलहरो गर्भसंधानस्थैर्यकृत् ।। (कै. नि.)

Śākaḥ kaṣāyāh śiśiro raktpịttapṛasādanaḥ
Kuṣthaśleṣmānilaharō gaṛbhasaṇdhānasthạịryakṛt.. (Kai. ni.)

Flower - Acrid, bitter, dry.

Bark - Acrid, sweet.

Wood - Acrid, cooling, laxative.

***Medicinal uses :***

Flower - Biliousness, bronchitis, urinary discharges.

Bark - Bronchitis.

Wood - Biliousness, dysentery, leucoderma, piles.

***Ethno- Medicinal uses :*** Stem bark - stomatitis (Joshi *et al.*, 1980); dysentery (Kataria & Gauchan, 2006); diarrhoea (Joshi *et al.*, 1980; Kataria & Gauchan, 2006).

Wood - Eye diseases (Sharma *et al.,* 1979); swelling (Janardhanan, 1963; Basak, 2006).

Fruits - Hair tonic (Sudhakar & Rolla, 1985).

Seed - Urinary problems (Rajwar, 1983).

## *TEPHROSIA PURPUREA* Pers. (PAPILIONACEAE)

***Syn.:*** *T. Villosa* Pers.; *T. maxima* Pers.; *T. lanceolata* R. Grah., *T. hamiltonaii* Drumm ex Gamble.

***Vern.:*** Sarphonka.

***Regional and Other Names :*** *Bengali* - Bon nil; *English* - Wild indigo; *Gujarati* – Ghodakah; *Hindi* - Jungli nil; *Kannada* - Empali, vajaraneeli; *Malayalam* - Kozhinijil, kaata miri; *Marathi* - Sharapunkha, udhadi; *Sanskrit* -Sharapunkhaa; *Tamil* - Kolingi; *Telugu* - Vempali, neelavempli.

***Distribution :*** All over India.

***Taxonomic description :*** Perennial, woody herbs, with 11-21 leaflets and reddish-purple or bright pink to white flowers in leaf-opposed racemes. Style glabrous. Pods linear.

***Flowering & Fruiting :*** July – February.

***Medicinal parts :*** Whole plant, root, Root bark, seeds.

***Taste :*** Bitter – astringent, acrid.

***Chemical constituents :*** Beta- Sitosterol, Cyanidin Chloride, Pongomol, Lupeol, Rutin, Dephinindinchloride, Karangin, Kangone, flevichapparins – B and C, Methylkaranjic acid, Purgurin.

Root contains Tephrosin, Degulin, Rotenone, Isotephrosin, alterative.

***Medicinal properties :***

शरपुंखो यकृतप्लीहगुल्मव्रणविषापहः ।<br>
तिक्तः कषायः कासास्त्रश्वासज्वरहरो लघुः ।। (भा.प्र.)

Śharapuṇkho yakṛitplīhguḷma bṛanabiṣāpahaḥ
Tiktaḥ kaṣāyaḥ kāsāṣṭraśvāsajvaraharo laghuḥ. (Bhā. Pra.)

Acrid, alexeteric, alterative anthelmintic, antipyretic, bitter, digestive.

***Medicinal uses :*** Diseases of blood, heart, liver, spleen. asthma, bronchitis, leprosy, snake bite, tumour, ulcer, wound.

***Ethno-Medicinal uses :*** Root-Stomatitis, colic, dysentery (Jain *et al.*, 1973); snake bite (Singh & Maheshwari, 1983); toothache (Sexena *et al.*, 1981); fever (Maheshwari *et al.*, 1986); stomach pain (Ganeshan *et al.*, 2006; Jadeja *et al.*, 2006; Rajendran & Rengamani, 2006).

Leaf - Wounds (Kiruba *et al.*, 2006); gas trouble (Arinathan *et al.*, 2006); leprosy (Sharma *et al.*, 2006).

Whole plant- diseases of heart and liver (Deokule, 2006).

Stem- Toothache, pyorrhoea (Aditya *et al.*, 2006).

## *TERMINALIA ARJUNA* (Roxb.) Wt. & Arn. (COMBRETACEAE)

***Vern.:*** Arjun,

***Regional and Other Names :*** *Bengali* – Arjun; *English* - Arjuna; *Gujarati* - Arunsadada; *Hindi* - Arjun; *Kannada* – Maddi; *Malayalam* – Vellilavu; *Marathi* – Arjuna; *Sanskrit* – Arjuna, Kukubha; *Tamil* - Vellaimarudamarum, vellamatta; *Telugu* - Tellamadoi, yerramaddi.

***Distribution :*** Throughout the Greater part of India.

***Taxonomic description :*** Trees, with smooth bark and leaves cordate at base. Flowers white, in axillary spikes or terminal panicles. Fruits 5-winged, wings curved upwards.

***Flowering & Fruiting :*** May – December,

***Medicinal parts :*** Bark, leaves, fruits.

***Taste :*** Astringent.

***Chemical constituents :*** Bark contains Bita- sitosterol, Arjunic acid, Friedlene, Glucoside, (Arjunetic), Tannins, Sugars, Sodium, Magnessium, Aluminium, Calcium Carbonate.

***Medicinal properties :***

ककुभः शीतलो हृद्यः क्षतक्षयविषास्त्रजित् ।
मेदोमेहव्रणान् हन्ति तुवरः कफपित्त हृत् ।। (भा.प्र.)

Kakubhaḥ śītalo hṛidyaḥ ksataksayaviṣāṣrajịt
Medomeha vṛaṇāṇ haṇti tuvaraḥ kaphapịtta hṛịt. (Bhā. Pra.)

Bark - Acrid, alexiteric, antidysenteric, cooling, heating, styptic, sweet, tonic.

***Medicinal uses :*** Bark - Asthma, biliousness, blood diseases, fractures, leucoderma, tumour, ulcer, urinary discharges.

***Ethno- Medicinal uses :*** Bark - Dysentery (Mahshwari *et al.,* 1986); heart diseases (Saxena & Vyas, 1983); dysentery (Jadhav, 2006).

Leaf - Wound (Sharma & Malhotra, 1984).

## ***TERMINALIA BELLERICA*** (Gaentn.) Roxb. (COMBRETACEAE)

***Vern.:*** Bahera,

***Regional and Other Names :*** *Bengali* – Bahera; *English* - Beleric myrobalan; *Gujarati* - Bahedo; *Hindi* - Bahera; *Malayalam* – Thanni; *Marathi* – Beheda; *Sanskrit* – Aksha, Bibhitaka; *Tamil* - Tanari; *Telugu* – Tandra,

***Distribution :*** Throughout the forests of India.

***Taxonomic description :*** Deciduous trees, with broadly obovate leaves clustered at the end of branches. Flowers yellow or creamy white, in spikes. Fruits ellipsoid, 5-ridged.

***Flowering & Fruiting :*** February – September,

***Medicinal parts :*** Fruits,

***Taste :*** Astringent, bitter.

***Chemical constituents :*** Fruits contain β-Sitosterol, gallic acid, ellagic acid, Chebulagic acid, Galloyl, Glucose, Many free sugars, Manitol, galactose, Rhamnose, glucoside.

***Medicinal properties :***

विभीतकं स्वादुपाकं कषायं कफपित्तनुत् ।
उष्णवीर्यं हिमस्पर्शं भेदनं कासनाशनम् ।। (भा.प्र.)

Vibhītakaṃ ṣwādupākaṃ kaṣāyaṃ kaphapiṭtanuṭ
Uṣṇavīryaṃ himaṣpaṛṣaṃ bhedanaṃ kāsanāsanaṃ. (Bhā. Pra.)

Fruits - Acrid, anthelmintic, bitter, digestive, laxative, Pungent.

***Medicinal uses :*** Bark- Anaemia, leucoderma

Fruits - Asthma, biliousness, bronchitis, cardiac diseases, inflammation, sore throat.

***Ethno- Medicinal uses :*** Seed - Gastric trouble (Singh & Pandey, 1980)

Bark - Cold (Jain & De, 1966).

Seed oil - Swelling, skin diseases (Kumar *et al.,* 2006).

Fruits - Diarrhoea, dropsy, fever (Bhalla *et al.,* 1982); stomachic (Tribedi *et al.,* 1982); liver complaint (Janardhanan, 1963); asthma, bronchitis (Shah *et al.,* 1983); diabetes (Tiwari *et al.,* 1979); leprosy (Pal & Jain, 1989; Basak, 2006); cough (Sharma & Malhotra, 1984); constipation (Roy & Chaturvedi, 1986); tonic, headache, diarrhoea (Nath *et al.,* 2006); swelling, skin diseases (Kumar *et al.,* 2006); malaria (Deka *et al.,* 2006); piles (Bhalla *et al.,* 1982; Basak, 2006).

## ***TERMINALIA CHEBULA*** (Gaentn.) Retz. (COMBRETACEAE)

***Vern.:*** Harra

***Regional and Other Names :*** *Bengali* – Haritaki; *English* - Black myrobalan; *Gujarati* - Hirdo; *Hindi* - Harra; *Malayalam* – Kadakai; *Marathi* – Hirda; *Sanskrit* – Abhaya, Haritaki; *Tamil* - Kadukkai; *Telugu* – Karakkai,

***Distribution :*** Throughout the Greater part of India.

***Taxonomic description :*** Deciduous trees, Leaves ovate or elliptic. Flowers Yellow or creamy white, in spikes. Fruits 5-ridged.

***Medicinal parts :*** Fruits.

***Taste :*** Astringent.

***Chemical constituents :*** Fruits contain- Tannic acid, gallic acid, chebulinic acid, mucilage, glycoside (Anthroquinine), Carbohydrates, amino acids, phosphoric acid, succinic acid.

***Medicinal properties :***

हरीतकी पंचरसाऽलवणा तुवरा परम् ।
रूक्षोष्णा दीपनी मेध्या स्वादुपाका रसायनी ।। (भा.प्र.)

Harītakī paṇcharasāalavaṇā tuvarā paraṃ
Rukṣoṣṇā dīpanī meḍhyā ṣvādupākā rasāyanī. (Bhā. Pra.)

Fruits - Alterative anthelmintic, antidysenteric, carminative, expectorant, heating, stomachic, tonic.

***Medicinal uses :*** Fruits - Anaemia, ascities, asthma, biliousness, constipation, elephantiasis, gout, hiccough, inflammation, leucoderma, piles, sore throat, strangury, tumour, typhoid, urinary discharges, vesicular calculi, vomiting.

***Ethno- Medicinal uses :*** Bark - Eczema (Purohit *et al.*, 1985).

Fruits - Sore throat (Sudhakar & Rolla, 1985); dysentery (Jain & De, 1966; Jadeja *et al.*, 2006); pneumonia (Roy & Chaturvedi, 1986); stomatitis (Vedprakash & Mehrotra, 1987); colic (Maheshwari *et al.*, 1986); eye diseases (Singh *et al.*, 1987); fever, cough, stomachache (Nath *et al.*, 2006; Baruah *et al.*, 2006); diarrhoea (Jadeja *et al.*, 2006); swelling, wounds, ulcers (Kumar *et al.*, 2006); malaria (Deka *et al.*, 2006); Gout (Jadeja *et al.*, 2006); small pox (Das *et al.*, 2006).

## *TINOSPORA CORDIFOLIA* (Willd.) Miers. ex Hk. f. & Th. (MENISPERMACEAE)

***Syn.:*** *T. glabra* (N. Burm.) Merr.; *Cocculus cordifolius* DC.; *Menispermum cordifolium* Willd.

***Vern.:*** Giloe.

***Regional and Other Names :*** *Bengali* – Golancha; *English* - Heart leaved moonseed,Gujarati-Gado; *Hindi* - Amrita; *Kannada* – Amrutaballi *Malayalam* - Amrytu *Marathi* - Gulvel *Sanskrit* – Amrita, Guduchi; *Tamil* - Shindil, kodi, amudam *Telugu* – Tippateege.

***Distribution :*** Throughout tropical India.

***Taxonomic description :*** Succulent, glabrous, deciduous climbers, with corky and lenticelled bark. Leaves ovate-cordate. Flowers yellow, in axillary racemes. Fruits globose, red when ripe.

***Flowering & Fruiting :*** December – June,

***Medicinal parts :*** Whole plant, Juice of fresh plant.

***Taste :*** Bitter.

***Chemical constituents :*** Starch, Glycoside (Giloin), Tinosporic acid, Vasa alcohol, essential oil, Vasa acid, Proteins, Calcium, Phosphorus, Berberine, gilosterol.

***Medicinal properties :***

गुडूची कटुका तिक्ता स्वादुपाका रसायनी ।
संग्राहिणी कषायोष्णा लघ्वी बल्याऽग्निदीपनी ।। (भा.प्र.)

Gudūchī katukā tiktā ṣvādupākā rasāyanī
Saṇgrāhiṇī kaṣāyoṣṇā laghvī bạlyāagnidīpanī. (Bhā. Pra.)

Stem - Bitter, diuretic, stimulant, stomachic.

***Medicinal uses :*** Stem - Burning sensation, constipation, diabetes, enriches the blood, fevers, Jaundice, skin diseases, spleen diseases tonic, vaginal discharges, vomiting.

***Ethno- Medicinal uses :*** Root - Diarrhoea, dysentery (Bhatt & Sabins, 1987); diabetes (Ghosh, 2006).

Leaf - Fever, headache, piles (Lakshamanan & Sankaranarayanan, 1988); joint pain (Ganeshan *et al.,* 2006); diabetes (Kadavul & Jisha, 2006); gastric trouble (Rajendran & Rengamani, 2006).

Whole plant - Bone fracture (Molla & Roy, 1985); piles (Singh, 1988); fever (Vedprakash & Mehortra, 1988); jaundice (Jadeja *et al.,* 2006).

Stem - Skin diseases, jaundice (Sharma *et al.,* 1979); stomachic, asthma (Shah & Joshi, 1971); abdominal gripes (Sarma *et al.,* 2006); diphtheria (Roy & Chaturvedi, 1987); Cough (Malhotra & Moorthy, 1973); leprosy & snake bite (Jadhav, 2006); joint diseases (Dwivedi *et al.,* 2005); fever (Kumar & Chauhan, 2005); jaundice (Deokule, 2006); dysentery (Kataria & Gauchan, 2006); malaria (Deka *et al.,* 2006); diarrhoea (Kataria & Gauchan, 2006; Gupta, 2006); ulcers, swelling (Das *et al.,* 2006).

## *TOONA CILIATA* Roem. (MELIACEAE)

***Syn.:*** *T. hexandra* (Wall. ex Roxb.) *Roemer.; Cedrala toona* Roxb.

***Vern.:*** Tuna.

***Regional and Other Names :*** *Bengali* – Lud; *English* - Red cedar; *Hindi* - Lud, Tuna; *Malayalam* – Akil; *Marathi* – Oeodari; *Sanskrit* - Apina; *Tamil* - Mali; *Telugu* – Gali.

***Distribution :*** Throughout India.

***Taxonomic description :*** A large deciduous tree; Leaves paripinnate; leaflets 8-30, lanceolate or ovate-lanceolate, entire or slightly undulate, often oblique at the base; flowers white, fragrant, in drooping, terminal panicles; capsule oblong, smooth, dark brown, septifragally 5-valved; seeds reddish-brown, winged at both ends.

***Flowering & Fruiting :*** March – June.

***Medicinal parts :*** Stem bark, flowers.

***Taste :*** Bitter.

***Chemical constituents :*** Wood & seeds contain A bitter substance-Nyctanthin, cedralone, β-sitosterol.

Flowers contain Baragapten flavone, glucoside, nyctanthin.

Bark contain Toonacilin, toonacilids, bitter, resin.

Wood contain Ferneslene, geronyl, geraniol, siderin, volatile oil.

***Medicinal properties :***

तूणी रक्तः कटुः पाके कषायो मधुरो लघुः ।
तिक्तो ग्राही हिमो वृष्यो व्रणकुष्ठास्त्रपित्तजित् ।। (भा. प्र.)

Tūṇī raktaḥ katuḥ pake kaṣayo madhurō laghuḥ
Tikto grahī himo vṛaṣyo vṛaṇkuṣthaṣṭrapiṭtajiṭ. (Bhā. Pra.)

Bark - Acrid, anthelmintic, aphrodisiac, astringent, bitter, digestive, expectorant, fattening, sweet, tonic.

***Medicinal uses :*** Bark - Biliousness, blood diseases, burning sensation, fever, headache, itching, leprosy, scabies, ulcer.

***Ethno- Medicinal uses :*** Leaf - Bronchitis (Molla & Roy, 1985).

Bark - Dysentery & fever (Sharma *et al.*, 1979).

## *TRIBULUS TERRESTRIS* L. (ZYGOPHYLLACEAE)

***Vern.:*** Gokhru.

***Regional and Other Names :*** *Bengali* – Gakhura; *English* - Land caltrops; *Gujarati* - Betagokhry; *Hindi* - Gokhru; *Kannada* – Negalu; *Malayalam* – Nerunji; *Marathi* – Lahangokhru; *Sanskrit* – Goksura, Bahukantaka; *Tamil* - Nerunji; *Telugu* – Chinnipalleru.

***Distribution :*** Throughout India.

***Taxonomic description :*** Procumbent annual herbs, with pinnate leaves and yellow flowers. Mericarps dorsally tuberculate, spines 4.

***Flowering & Fruiting :*** March – November.

***Medicinal parts :*** Fruits, roots.

***Taste :*** Sweet.

***Chemical constituents :*** Fruits contains Alkaloid, fixed oil, essential oil, Resin, Steroids, Saponins, Vitamin- C, Protein, Nitrates.

***Medicinal properties :***

गोक्षुरः शीतलः स्वादुर्बलकृद् वस्तिशोधनः ।
मधुरो दीपनो वृष्यः पुष्टिदश्चाश्मरीहरः ।। (भा.प्र.)

Gokṣuraḥ śītalaḥ ṣvāduṛbalakṛiḍ vaṣtiśodhnaḥ
Madhurō dīpano vṛiṣyaḥ puṣtidaścāśmarīharaḥ. (Bhā. Pra.)

Root - Alterative, aphrodisiac, appetiser, cooling, fattening, Sweet, tonic.

Fruits - Alterative, aphrodisiac, appetiser, cooling, fattening, Sweet, tonic.

Seed - Aphrodisiac, cooling, diuretic, fattening.

***Medicinal uses :*** Root - Asthma, blood diseases, burning sensation, cough, inflammation, pruritus, strangury urinary discharges, vesicular calculi.

Fruits - Asthma, blood diseases, burning sensation, cough, inflammation, pruritus, strangury urinary discharges, vesicular calculi.

Seed - Inflammation, stones in bladder, urinary trouble.

***Ethno-Medicinal uses :*** Fruits - Leucorrhoea (Saxena & Vyas, 1983); cough (Malhotra & Moorthy,1973); asthma (Singh & Pandey, 1980); urinary complaint (Shah *et al.,* 1981; Sharma *et al.,* 2006); liver complaint (Gopal & Shah, 1985); kidney diseases (Deokule, 2006).

Whole plant - Impotency (Shekhawat & Anand, 1984); leucorrhoea (Jadhav, 2006).

Leaf - Jaundice (Ganeshan *et al.,* 2006); gonorrhoea, rheumatism (Shah, 2005); Gout (Jadeja *et al.,* 2006).

## ***URGINEA INDICA*** (Roxb.) Kunth. (LILIACEAE)

***Syn.:*** *Drimia indica* (Roxb.) Jessop.

***Vern.:*** Jangli kanda.

***Regional and Other Names :*** *Bengali* – Banpiaaj; *English* - Indian squill; *Gujarati* - Rankando; *Hindi* - Jangli piyaz; *Malayalam* – Kattulli; *Marathi* – Janglipyajha; *Sanskrit* - Kolakanda; *Tamil* - Narivaengayam; *Telugu* - Adavi thellagadda.

***Distribution :*** W. Himalaya, Bihar, Chota Nagpur.

***Taxonomic description :*** A small geophyte with ovoid, tunicated, white bulb, leaves radical, linear-lanceolate; flowers pale-brown, on a leafless scape, in few flowered racemes; capsule long, ellipsoid; seeds elliptic, black.

***Medicinal parts :*** Bulbs.

***Taste :*** Bitter-Acrid.

***Chemical constituents :*** Bulb contain Glucosides, Scillian, Scillipicrin, Scillitin, Mucilage, Sugars, Calcium oxalate.

***Medicinal properties :***

कोलकन्दः कटुश्चोष्णः कृमिकासकफापहः ।
उत्क्लेशवान्तिजननो हृदयः श्वासनिवारणः ।। (द्र. गु. वि.)

Kolakaṇdaḥ katuśchoṣṇaḥ kṛimikāskaphāpaḥ
Uṭkleshshvāṇtijannano hṛidayaḥ śvāsnivāranaḥ. (Dra. Vi.)

Anthelmintic, cardio tonic, diphoretic, diuretic, expectorant, laxative.

***Medicinal uses :*** Asthma, blood pressure, cough, menorrhagia, paralysis, pneumonia fever, skin diseases, whooping cough.

***Ethno- Medicinal uses :*** Bulb - Abortifacient (Tarafder, 1983); skin dieseases (Malhotra & Moorthy, 1973; Sarma *et al.,* 1979); fever (Tarafder, 1986); wounds (Uniyal & Chauhan, 1973); rheumatism (Saxena, 1986).

Leaf - Bone fracture (Ganeshan *et al.,* 2006).

## *VERNONIA CINEREA* (L.) Less (ASTERACEAE)

***Syn.:*** *Conyza cinerea* L.

***Vern.:*** Sahadevi.

***Regional and Other Names :*** *Bengali* - Kukushim, Kalajira; *English* - Purple fleabane; *Gujarati* - Sadodi; *Hindi* - Sahadevi; *Kannada* – Sahadevi; *Malayalam* - Poovankurunthila, Puvankadanthel; *Marathi* - Sahadivi, Osari, Sadodi; *Sanskrit* - Sahadevi; *Tamil* - Puvamkurundal, Mukuttipundu; *Telugu* - Garitikamma, garitikamini.

***Distribution :*** Throughout India.

***Taxonomic description :*** Erect or decumbent herbs, with ovate to obovate or variable leaves. Heads violet or pink. Achenes terete.

***Flowering & Fruiting :*** Throughout the year.

***Medicinal parts :*** Root, flower, seeds, whole plant.

***Taste :*** Bitter.

***Chemical constituents :*** Plant contains Lupeol acetate, alpha – amyrin, acetate, glucoside, Quercetin, β- sitosterol, stigma sterol, œ- spina sterol, potassium chhoride, œ- amyrin, lupeol.

Flower contains Luteolin, Iso – orientin, chrysoeriol.

***Medicinal properties :***

ज्वरं हन्ति शिरोबद्धा सहदेवीजटा यथा ।। (च. सू.)

Jvaraṃ haṇti śirōbaḍdh ā sahadevījatā yathā. (Ca. su.)

Plant - Astringent, cold, stomachic, sweet, tonic.

***Medicinal uses :*** Plant - Asthma, bronchitis.

***Ethno- Medicinal uses :*** Seed - Skin diseases (Bhalla *et al.,* 1982).

Root - Leucorrhoea (Saxena & Vyas, 1983).

Whole plant - Wound (Malhotra & Moorthy, 1973); piles (Rajwar, 1983); fever (Bedi, 1978); dyspepsia (Jadeja *et al.,* 2006); eczema, ringworm, elephantiasis (Basak, 2006).

Leaf - Malaria (Ramachandran & Nair, 1981); piles (Deokule, 2006); constipation (Singh *et al.*, 1987); diarrhoea, leucorrhoea (Das *et al.,* 2006).

## *VITEX NEGUNDO* L. (VERBENACEAE)

***Vern.:*** Nirgundi.

***Regional and Other Names :*** *Bengali* - Nisinda, samalu; *English* - Chaste tree, Indian private; *Gujarati* - Nagaol; *Hindi* - Shamalu, sambhalu; *Kannada* - Zakkigida, nekki; *Malayalam* - Indrani, vellanocchi; *Marathi* – Nirgundi; *Sanskrit* - Nirgudi; *Tamil* - Vellai nocohi, nirkkundi; *Telugu* - Nallavavilli, vaavilli.

***Distribution :*** Throughout India.

***Taxonomic description :*** Evergreen shrubs, with 3 to 5-foliolate leaves. Flowers violet, in peduncled cymes. Fruits ovoid, 4-celled.

***Flowering & Fruiting :*** September – April.

***Medicinal parts :*** Whole plant, leaves, fruits, seeds.

***Taste :*** Bitter, acrid.

***Chemical constituents :*** Volatile oil, Alkaloids (Nergundin & Hydrocotylon), Vitamin- C, Carotene, flavon, Tannic acid, Hydroxy benzoic acid, Resin, Mallic acid.

***Medicinal properties :***

निर्गुण्डी कटुतिक्तोष्णा कृमिकुष्ठ रुजापहा ।
वातश्लेष्मप्रशमनी प्लीहा गुल्मारूचीर्जयेत् ।। (ध.नि.)

Nirgundī karutiktoṣhṇā kṛimikuṣtha rujāpahā
Vātaśleṣmapraśmani plīhā gulīma ruchīrjayet. (Dha. Ni.)

Acrid, anthelmintic, astringent, bitter, cephalic, heating, Pungent, stomachic.

***Medicinal uses :*** Asthma, biliousness, bronchitis, enlargement of the spleen, eye diseases, inflammation, leucoderma.

***Ethno- Medicinal uses :*** Root - Piles (Singh, 1988); rheumatism (Gupta, 1981); diarrhoea, dysentery (Kataria & Gauchan, 2006).

Seed - Gout (Singh & Pandey, 1980).

Flower - Pneumonia (Virjee *et al.*, 1984); diarrhoea (Malhotra & Moorthy, 1973); dysentery (Sharma *et al.*, 1979); eczema, syphilis, skin diseases, leprosy (Purohit *et al.*, 1985).

Stem - Bone fracture (Gupta, 1981); fever (Tribedi *et al.*, 1985); sciatica (Hembrom & Goel, 2005).

Leaf - Boils, blister (Maheshwari & Singh, 1984); paralysis (Nagendra Prasad & Abraham, 1984); cold & cough (Jain & De, 1966); colic (Rajwar, 1983); bodyache (Das *et al.*, 1983); gum trouble, Icthing (Tarafder, 1978); ulcer (Bhalla *et al.*, 1982); wound (Tarafder, 1984); swelling (Rao, & Jamir, 1982; Deokule, 2006); fever (Shah *et al.*, 1983); headache (Shah, 2005); sinusitis (Chhetri, 2005; Rajendran & Rengamani, 2006); jaundice (Purkayastha & Subhan, 2006; Jadeja *et al.*, 2006); joint diseases (Dwivedi *et al.*, 2005); bodyache, muscular pain (Pereira, 2006; Deokule, 2006); rheumatism (Tarafder, 1978; Mokat & Deokule, 2006; Das *et al.*, 2006); malaria (Deka *et al.*, 2006); Gout (Jadeja *et al.*, 2006); migrain (Sharma *et al.*, 2006).

Whole plant – Asthma, lung diseases, tonsilitis, rheumatism, urinary troubles (Kumar *et al.*, 2006).

## *WITHANIA SOMNIFERA* (L.) Dunal (SOLANACEAE)

***Vern.:*** Ashwagandha.

***Regional and Other Names :*** *Bengali* – Ashvaganda; *Gujarati* - Asan; *Hindi* - Ashvagandha; *Kannada* – Viremaddlinagadde; *Malayalam* – Pevetti; *Marathi* – Askondhatilli; *Sanskrit* - Ashvagandha; *Tamil* - Amukkura; *Telugu* - Putivendram, pannerugadda.

***Distribution :*** In the drier part of India.

***Taxonomic description :*** Hoary-tomentose undershrubs, with elliptic-lanceolate or obovate leaves, bisexual flowers and berries enclosed in the persistent calyx.

***Flowering & Fruiting :*** August- December.

***Medicinal parts :*** Root, leaves, fruits, seeds, whole plant.

***Taste :*** Bitter, acrid.

***Chemical constituents :*** Root contains Alkaloids (Somniferon), Withami-non, Wasamin, Sugars, Glycosides, Amino acids, essential oils, Withaniol, Hentriacontane, Phytosterol & Oils.

Leaves contain Withanoloids alkaloids, glycosoids, glucose, amina acids.

Stem contains Proteins, amino acids, Tannin, flavonoids, Calcium, Iron.

***Medicinal properties :***

अश्वगन्धाऽनिलश्लेष्मश्वित्र शोथक्षयापहा ।
बल्या रसायनी तिक्ता कषायोष्णाऽतिशुक्रला ।। (भा.प्र.)

Aśvagaṇdhāanilaśleṣmaśviṭra śothakṣhyāpahā
Baḷyā rasāyanī tiktā kaṣāyoṣṇāatiśukṛalā. (Bhā. Pra.)

Tuber - Acrid, alterative, anthelmintic, aphrodisiac, bitter, tonic.

***Medicinal uses :*** Tuber - Asthma, bronchitis, inflammation, insomnia, psoriasis, scabies, ulcers.

***Ethno- Medicinal uses :*** Seed - Chest complaint (Bhalla *et al.,* 1982).

Leaf - Boils, eye complaint (Bhalla *et al.,* 1982); skin diseases (Lal & Yadav, 1983).

Root - Cough, dropsy, dyspepsia, rheumatism, (Bhlla *et al.,* 1982); asthma, bronchitis (Shah *et al.,* 1983; Sharma *et al.,* 2006); wound (Tarafder, 1984); epilepsy (Tiwari *et al.,* 1979); general debility, spermatorrhoea (Kumar & Chauhan, 2005); tonic (Deokule, 2006); sexual vigour (Mokat & Deokule, 2006; Chatterjee & Bhatterjee, 2006); diabetes (Ghosh, 2006).

## *WOODFORDIA FRUTICOSA* (L.) Kurz. (LYTHRACEAE)

***Syn.:*** *W. floribunda* Salisb.

***Vern.:*** Dhai.

***Regional and Other Names :*** *Bengali* – Dhatriful; *English* - Fire flame; *Gujarati* - Dhawadina; *Hindi* - Dhai, dhawi; *Kannada* – Tamrapusphi; *Malayalam* – Gaddaisinka; *Marathi* - Phulsatti, dhaiphal; *Sanskrit*- Dhataki; *Tamil* - Velakkai; *Telugu* - Sirinji, jargi.

***Distribution :*** Throughout India.

***Taxonomic description :*** Shrubs, with broadly lanceolate leaves and deep-red flowers in axillary cymes. Capsules oblong.

***Flowering & Fruiting :*** September– February.

***Medicinal parts :*** Bark, Flower.

***Taste :*** Acrid.

***Chemical constituents :*** Leaves contain Ellagic acid & polyphenols, lawsone & tannins.

Stem contain Octacosnol, β-sitosterol.

Flowers contain Hecogenin & gallic acid, tannins, soluble nontans & sugars.

***Medicinal properties :***

धातकी कटुका शीता मृदुकृत्तुवरा लघुः ।
तृष्णाऽतीसारपित्तास्त्रविषक्रिमिविसर्पजित् ।। (भा. प्र.)

Dhātkī katukā śītā mṛadukṛuttuvarā laghuḥ
Tṛaṣḥṇāatīsārpịttāṣṭraviṣhkṛimivisaṛpjịt. (Bhā. Pra.)

Bark - Acrid, alexiteric, anthelmintic, cooling, Pungent, tonic.

***Medicinal uses :*** Bark-Blood diseases, dyspepsia, erysipelas, leprosy.

Flowers - Leucorrhoea, menorrhagia, toothache.

***Ethno- Medicinal uses :*** Root - Rheumatism (Jain & De, 1966).

Stem bark - Burns (Jain, 1965); diarrhoea, dysentery (Kadavul & Jisha, 2006).

Leaf - Injury (Maheshwari & Singh, 1987).

Fruits - Dysentery (Singh & Pandey, 1980).

Flowers - Injury (Sharma & Malhotra, 1984); small pox (Bhalla *et al.*, 1982); menorrhoea (Saxena & Vyas, 1983); cough (Gupta, 1981); dysentery (Chhetri, 2005; Jadeja *et al.*, 2006); diarrhoea (Chhetri, 2005); miscarriage (Hembrom & Goel, 2005).

## *XANTHIUM STRUMARIUM* L. (ASTERACEAE)

***Syn.:*** *X. indicum* Koenig ex Roxb.

***Vern.:*** Adhasisi.

***Regional and Other Names :*** *Bengali* – Chotadhatura; *English* - Burweed; *Gujarati* - Gadriyun; *Hindi* - Banokra; *Malayalam* – Maruluumatti; *Marathi* – Dutuudi; *Sanskrit* - Arishta; *Tamil* - Maruloomadtam; *Telugu* - Marula, matangi.

***Distribution :*** Throughout India.

***Taxonomic description :*** Foetid-smelling unarmed herbs, about 1m high. Leaves ovate-suborbicular. Heads greenish yellow, in terminal and axillary racemes.

***Flowering & Fruiting :*** January – February.

***Medicinal parts :*** Whole plant.

***Taste :*** Acrid.

***Chemical constituents:***

Seed contains Xanthostrumarin acid, oxalic acid.

Fruits contain Strumaroside, stigmasterol, xanathin, xanthinin, xanthumin, xanthanol.

***Medicinal properties:*** Acrid, alexiteric, alterative, anthelmintic, antipyretic, appetiser, cooling, digestive, fattening, laxative, tonic.

***Medicinal uses:*** biliousness, epilepsy, fever, leucoderma, salivation.

***Ethno- Medicinal uses :*** Seed - Wound (Pal & Jain, 1989); headache (Maheshwari *et al.*, 1986).

Fruit - Eye diseases (Maheshwari & Singh, 1987); small pox (Deokule, 2006); rheumatism (Das *et al.*, 2006).

Leaf - Malaria (Pal & Jain, 1989); ringworm (Maheshwari & Singh, 1987).

Flower - Toothache (Sudhakar & Rolla, 1985).

Root - Ulcer (Malhotra & Moorthy, 1973); snake bite (Kumar & Chauhan, 2005).

## *ZINGIBER OFFICINALE* Rosc. (ZINGIBERACEAE)

***Vern.:*** Adrak.

***Regional and Other Names :*** *Bengali* – Ada; *English* - Ginge;r *Hindi* - Ada; *Kannada* – Shunti; *Malayalam* – Inchi; *Marathi* – Ale; *Sanskrit* - Ardraka; *Tamil* - Allam; *Telugu* – Allamu.

***Distribution :*** Widely Cultivated eleswhere.

***Taxonomic description :*** A small herb with horizontal, fleshy rhizome; leaves distichous, narrow, oblong-lanceolate; Flowers on long spike and are of purple colour on spike, small.

***Medicinal parts :*** Rhizome.

***Taste :*** Acrid pungent.

***Chemical constituents :*** Rhizome contains Essential oil, Fibres, Proteins, Starch, Carbohydrates, Resin, Glutamin, Thrionin, Free amino acids, Zingiberol, Zingiberin, Glutamic acid, Aspartic acid.

***Medicinal properties :***

शुण्ठी रूच्यामवातघ्नी पाचनी कटुका लघुः ।
स्निग्धोष्णा मधुरा पाके कफवात्विबन्धनुत् ।। (भा.प्र.)

Śuṇthī rucyāmvātaghnī pācnī katukā laghuḥ
Ṣnigdhoṣṇā madhurā pāke kaphvāṭvibaṇdhanuṭ. (Bhā. Pra.)

Rhizome - Analgesic, aphrodisiac, appitizer, aromatic, blood purifier, carminative, digestive, expectorant, febrifuge stimulant, thermogenic.

***Medicinal uses :*** Rhizome - Arthritis, asthma, chronic bronchitis, cold, cough, dysuria, flatulence, gout, joint problems, lumbago, muscular trouble, rheumatism, sciatica, slip disc.

***Ethno-Medicinal uses :*** Rhizome - Cough, Insect sting, Snake bite (Kharkongor & Joshep, 1981); cough, cold (Pereira, 2006; Deokule, 2006); Abortifacient (Tarefder, 1953); Post delivery period (Jadhav, 2006); headache (Deokule, 2006); gastric trouble (Gupta, 2006); constipation, dyspepsia (Jadeja *et al.,* 2006); malaria (Deka *et al.,* 2006); Gout (Jadeja *et al.,* 2006).

## *ZIZIPHUS MAURITIANA* Lam. (RHAMNACEAE)

***Syn.:*** *Z. jujuba* Mill.; *Z. vulgaris* Lam.; *Z. sativa* Gaertn.

***Vern.:*** Ber.

***Regional and Other Names :*** *Bengali* - Kul, borui; *English* - Indian plum; *Gujarati* - Bor; *Hindi* - Ber; *Kannada* – Yelachi; *Malayalam* – Elantha; *Marathi* – Bor; *Sanskrit* - Badari; *Tamil* - Elantharandor; *Telugu* - Regu, pandu, pazham.

***Distribution :*** Throughout India.

***Taxonomic description :*** Evergreen small trees or shrubs, with droopling branches armed with prickles. Leaves ovate-elliptic to suborbicular, tomentose beneath. Flowers greenish-yellow, in axillary cllusters. Drupes fleshy.

***Flowering & Fruiting :*** March – May.

***Medicinal parts :*** Bark, leaves, fruits.

***Taste :*** Sweet.

***Chemical constituents :*** Fruits contains Sugars & mucilage.

***Medicinal properties :***

सौवीरं बदरं शीतं भेदनं गुरु शुक्रलम् ।। (भा.प्र.)

Saūvīrum badram śītam bhednam guru śukralam. (Bhā. Pra.)

Root - Bitter, cooling.

Leaf - Antipyretic, bitter, cooling.

Fruits - Aphrodisiac, cooling, indigestible, laxative, tonic.

Seed - Aphrodisiac, acrid, sweet.

***Medicinal uses :*** Root - Biliousness, headache.

Bark - Boils, diarrhoea, dysentery.

Leaf - Biliousness, diarrhoea, obesity.

Fruits - Biliousness, blood diseases, vomiting.

Seed - Asthma, burning sensation, cough, eye diseases, leucorrhoea, vomiting.

***Ethno-Medicinal uses :*** Fruits - Blood purifier (Bhalla *et al.,* 1982); stomachache (Kharokongor & Joseph, 1981; Baruah *et al.,* 2006); Hepatitis (Dutta Choudhury & Shil, 2006).

Bark - Diarrhoea (Joshi *et al.,* 1980; Kataria & Gauchan, 2006); piles (Deokule, 2006); dysentery (Kataria & Gauchan, 2006).

Leaf - Rheumatism (Dar *et al.,* 1984); whooping cough (Virjee *et al.,* 1984); conjuctivitis (Bhalla *et al.,* 1982); scorpion sting (Shah *et al.,* 1981); dizziness (Sarma *et al.,* 2006); Hepatitis (Dutta Choudhury & Shil, 2006).

Root - Stomachic (Tarafder, 1983); ulcer, wound (Janardhanan, 1963); dysentery (Jadhav, 2006; (Basak, 2006).

# GLOSSARY OF MEDICAL TERMS

| | |
|---|---|
| *Abortifacient* | an agent that induces abortion |
| *Acrid* | biting, pungent |
| *Adenitis* | inflammation of a gland or group of glands (or glandlike structures) |
| *Agalactia* | absence or failure of secretion of milk |
| *Alexipharmic* | antidote to poison |
| *Alexiteric* | protective to infectious diseases |
| *Alopecia* | loss of hair : a malady in which the hair falls from one or more circumscribed round or oval areas, leaving the skin smooth and white |
| *Alterative* | causing a favourable change in the disordered functions of the body or metabolism |
| *Amenorrhoea* | failure of menstruation |
| *Amentia* | an arrest of the development of the mind from birth to early age |
| *Anaemia* | lack of enough blood causing paleness |
| *Analgesic* | an anodyne |
| *Anaphrodisiac* | having the power to lessen or inhibit sexual feeling |
| *Anodyne* | a medicine that allays pain |
| *Anorexia* | a condition of having lost the appetite for food |
| *Anthelmintic* | destroying or expelling worms |
| *Antibacterial* | active against bacteria |
| *Antidote* | an agent which neutralises or opposes the action of a poison |
| *Anti-inflammatory* | acting against inflammation |
| *Anti-leprotic* | any agent which prevents or cures leprosy |
| *Antilithic* | an agent which prevents the formation of calculi |
| *Anti-periodic* | preventing the regular recurrence of a disease |
| *Antipyretic* | counteracting fever |
| *Antiscorbutic* | acting against scurvy |
| *Anti-septic* | a chemical sterilising substance to kill or control pathogenic microbes |
| *Antispasmodic* | opposing Spasms |
| *Anuria* | failure of the kidneys to produce urine |

| | |
|---|---|
| *Aperient* | A laxative |
| *Aphrodisiac* | drug which stimulates sexual desire |
| *Areata* | ocurring in circumscribed areas or patches |
| *Arthritis* | inflammation of joints |
| *Ascities* | an abnormal accumulation of a fluid in the peritoneal cavity |
| *Astringent* | a drug which contracts the organic tissues and canals of the body, and therby checking or deminishing bleeding of excessive discharges |
| *Azoospermia* | the comlete absence of sperm from the seminal fluid |
| *Baldness* | absence of hair from areas where it normally grows |
| *Beriberi* | a deficiency disease caused by imbalance of carbohydrate and vitamin B |
| *Biliousness* | a symptom complex comprising nausea |
| *Boil* | tender inflamed area of the skin containing pus |
| *Bronchitis* | inflammation of the bronchi |
| *Bronchodilator* | an agent that causes widening of the air passages by relaxing bronchial smooth muscle |
| *Bubo* | an inflammatory swelling of a lymph gland |
| *Calculi* | a hard pebble-like mass formed within the body, particu-larly in the gall bladder or anywhere in the urinary tract |
| *Cancer* | the abnormal and uncontrolled division of cells that then invade and destroy the surrounding tissues |
| *Carbuncle* | an infection of the skin and subcutaneous tissue by Staphylococcus aureus |
| *Carminative* | drug curing flatulence |
| *Catarrh* | inflammation of a mucous membrane |
| *Cathartic* | having the power of cleansing the bowels- purgative |
| *Cephalagia* | headache |
| *Cephalic* | a remedy for disorders of the head |
| *Cholera* | a severe infectious epidemic disease due to Vibrio cholerae |
| *Chronic bronchitis* | the patient coughs up excessive mucus secreted by enlarged bronchial mucous glands |
| *Conjunctivitis* | inflammation of conjunctiva |
| *Colic* | a severe spasmodic griping pain |
| *Consumption* | pulmonary tuberculosis |
| *Contraceptive* | drug which prevents unwanted pregnancy |
| *Convulsion* | a violent involuntry contraction of the skeletal musculature |
| *Cystitis* | inflammation of a bladder, especially the urinary bladder |
| *Dandruff* | dead scarf-skin seperating in small scales and entangled in the hair |
| *Deafness* | partial or total loss of hearing in one or both ears |
| *Demulcent* | soothing |
| *Depurative* | an agent that purifies the blood |

| | |
|---|---|
| *Diabetes* | here it refers to diabetes mellitus: a disorder of carbohy-drate metabolism in which sugars in the body are not oxidized to produce energy due to lack of the pancreatic hormone-insulin. The accumulation of sugar leads to its appearance in the blood |
| *Diaphoretic* | a drug which induces perspiration |
| *Diarrhoea* | frequent bowel evacuation or the passage of abnormally soft or liquid faeces |
| *Diptheria* | a specific infectious disease caused by virulent strains of a bacillus |
| *Diuretic* | promoting the discharge of urine |
| *Dropsy* | an excessive accumulation of a clear or watery fluid in any of the tissues or cavities of the body |
| *Dysentery* | an infection of the intestinal tract causing severe diarrhoea with blood and mucus |
| *Dysmenorrhoea* | painful menstruation |
| *Dysopia* | defective vision |
| *Dyspepsia* | indigestion |
| *Dysuria* | difficulty or pain while passing urine |
| *Earache* | pain in the ear |
| *Eczema* | a noncontagious inflammatory disease of the skin with much itching and burning |
| *Elephantiasis* | gross lymphatic oedema of the limbs leading to hypertrophy |
| *Emetic* | causing vomiting |
| *Emollient* | softening |
| *Epilepsy* | an affection of the nervous system resulting from excessive or disordered discharge of cerebral neurons |
| *Eruptions* | the outbreak of a rash |
| *Erysipelsa* | an inflammatory disease generally on the face marked by a bright redness of the skin |
| *Expectorant* | aiding the secretion of the mucous membrane of the air passages and the removal of fluid by spitting |
| *Fatigue* | mental or physical tiredness, following prolonged or intense activity |
| *Febrifuge* | that which drives off fever |
| *Fever* | a rise in body temperature above the normal |
| *Fistula* | an open channel from the anus or rectum to the skin near the anus |
| *Flatulence* | distention of stomach |
| *Foetid* | bad-smelling, fetid |
| *Gleet* | chronic discharge of thin mucus from the vagina |
| *Goitre* | enlargement of the thyroid gland |
| *Gonorrhoea* | an inflammatory disease of the urinogenital passagaes characterised by pain and discharge |
| *Gout* | a disease of purine metabolism characterised by attacks of arthritis with an associated raised level of serum uric acid |

| | |
|---|---|
| *Haematemesis* | vomitting of blood |
| *Haematinic* | any substance which is required for the production of RBCs and its constituents |
| *Haemoptysis* | spitting of blood |
| *Haemorrhage* | the escape of blood from a ruptured blood vessel, externally or internally |
| *Haemorrhagic diseases* | associated with or resulting from blood loss |
| *Haemostatic* | styptic |
| *Hallucination* | a false perception of something that is not really there |
| *Headache* | pain felt deep within the skull |
| *Helminthiasis* | morbid state due to infestation with worms |
| *Helminthic* | Anthelmintic, vermifugal |
| *Hemicrania* | headache confined to one side |
| *Hepatomegaly* | enlargement of the liver |
| *Hiccough* | abrupt involuntary lowering of the diaphragm and closure of the sound-producing folds at the upper end of the trachea, producing a characteristic sound as the breath is drawn in |
| *Hydrocele* | a circumscribed collection of fluid in the tunica vaginalis testis |
| *Hydrophobia* | exaggerated fear of water as in rabies |
| *Hypertension* | high arterial blood pressure |
| *Hysteria* | a neurotic disorder with varying symptoms |
| *Impotency* | inability to perform the sexual act due to failure of the reflex mechanism |
| *Inflammation* | the body's response to injury, which may be acute or chronic |
| *Insecticidal* | any agent which kills or destroys insects |
| *Insomnia* | the condition a being unable to sleep/ |
| *Intermittient fever* | a fever that rises, subsides, then returns again |
| *Itcing* | discomfort or irritation of the skin, prompting the sufferer to scratch or rub the affected area |
| *Jaundice* | a yellowing of the skin or whites of the eyes, indicating excess bilirubin (a bile pigment) in the blood |
| *Laxative* | a drug used to stimulate or increase the frequency of bowel evacuation, or to encourage the passage of a softer or bulkier stool |
| *Leprosy* | a chronic disease, caused by the bacterium Mycobacterium leprae, that affects the skin, mucous membranes, and nerves |
| *Leucoderma* | loss of pigments in areas of the skin, resulting in the appearance of white patches or bands |
| *Leucoma* | white opacity occurring in the cornea |
| *Leucorrhoea* | whitish or yellowish discharge of mucus from the vaginal opening |
| *Lumbago* | pain in lower back |
| *Mammillitis* | inflammation of the nipple |

| | |
|---|---|
| *Menorrhagia* | excessive or prolonged menstruation |
| *Migrain* | condition resulting from spasm and subsequent overdila-tation of certain arteries in the brain, which causes a recur-rent throbbing headache that characteristically affects one side of the head |
| *Miscarriages* | the expulsion or removal of an embryo or fetus from the uterus at a stage of pregnancy when it is incapable of independent survival |
| *Mumps* | an acute infectious disease caused by a virus |
| *Nasitis* | inflammation of the nose |
| *Nightblindness* | the inability to see in dim light or at night |
| *Obesity* | an excessive accumulation of fat in the body |
| *Oedema* | excessive accumulation of fluid in the body tissues |
| *Oleaginous* | oily |
| *Ophthalmia* | a term usually applied to conjunctivitis |
| *Ophthalmopathy* | any disease of the eye |
| *Otalgia* | pain in the ear |
| *Otorrhoea* | any discharge from the ear, commonly a purulent discharge in chronic middle ear infection |
| *Paralysis* | a muscle weakness that varies in its extent, its severity, and the degree of spasticity or flaccidity according to the nature of the underlying disease and its distribution in the brain, spinal cord, peripheral nerves, or muscles |
| *Pectoral diseases* | effective in diseases of the chest |
| *Peptic ulcer* | a breach in the lining (mucosa) of the digestive tract produced by digestion of the mucosa by pepsin and acid |
| *Piles* | enlargement of the normal spongy blood-filled cushions in the wall of the anus, usually a consequence of prolonged constipation or ocassionally, diarrhoea |
| *Pneumonia* | a general disease in which the essential lesion is an inflammation of the spongy tissue of the lung with consolidation of the alveolar exudate |
| *Pneumonosis* | any lung disorder |
| *Polyuria* | the production of large volumes of urine, which is dilute and of a pale colour |
| *Procreant* | the drug which begets |
| *Prophylactic* | pertaining to the prevention of the development of a disease |
| *Prurigo* | an eruption of the skin causing severe itching |
| *Pruritus* | itching |
| *Pyorrhoea* | disease of the tissues that support and attach the teeth-the gums, periodontal membrane, and alveolar bone |
| *Refrigerant* | cooling |
| *Remittent fever* | a remittent fever is one in which body temperature flactuates but does not return to normal |
| *Renal calculi* | calculi relating to kidney |

| | |
|---|---|
| *Resolvent* | causing resolution of a tumour or swelling |
| *Retention of urine* | inability to pass urine, which is retained in the bladder |
| *Revulsive* | causing revulsion in drawing away of blood from a pathological area to another area |
| *Rheumatalgia* | rheumatic pain |
| *Rheumatism* | any disorder in which aches and pains affect the musclas and joints |
| *Rheumatoid arthritis* | it is the second most common form of arthritis. It typically involves the joints of the fingers, wrists, feet and ankles with later involvement of the hips, keens, shoulders and neck |
| *Rhinitis* | inflammation of the nasal mucous membrane |
| *Ringworm* | fungus infection of the skin, the scalp, or the nails |
| *Rubefacient* | reddening of the skin |
| *Scabies* | sarcoptic infestation of the human skin particularly a contagious skin disease caused by invasion of the epidermis |
| *Sciatica* | pain felt down the back and outer side of the thigh, leg, and foot |
| *Scrofula-* | tuberculous cervical adenitis with or without ulceration |
| *Scurvy* | a deficiency disease due to lack of vitamin C |
| *Sinusitis* | inflammation of one or more of the mucous- membrane-lined air spaces in the facial bones that communicate with the nose |
| *Slipdisc* | a colloquial term for a prolapsed intervertebral disc |
| *Small Pox* | an acute infectious virus disease causing high fever and a rash that scars the skin |
| *Somatalgia* | pain in the body |
| *Soporific* | inducing sleep |
| *Sore throat* | pain at the back of the mouth, commonly due to bacterial or viral infection of the tonsils |
| *Spasmolytic* | a drug that relieves spasm of smooth muscle |
| *Splenopathy* | any disease of the spleen |
| *Spermatorrhoea* | the involuntary discharge of semen without orgasm |
| *Stomachache* | pain in the stomach |
| *Stomachic* | a medicine that increases the functional activity of the stomach |
| *Stomatitis* | generalised inflammation of the oral mucosa |
| *Strangury* | severe pain in the urethra referred from the base of the bladder and associated with an intense desire to pass urine |
| *Styptic* | having the power to arrest bleeding |
| *Sudorific* | an agent which induces sweating |
| *Sun stroke* | raised body temperature (pyrexia), absence of sweating, and eventual loss of consciousness due to failure or exhaustion of the temperature-regulating mechanism of the body |
| *Syphilis* | contagious venereal disease |
| *Tetanus* | an infective disease due to toxin of Clostridium tetani |

| | |
|---|---|
| *Tuberculosis* | an infectious disease caused by the bacillus Mycobacterium tuberculosis and characterized by the formation of nodular lesions in the tissues |
| *Tumour* | any abnormal swelling in or on a part of the body |
| *Typhoid* | an infection of the digestive system by the bacterium Salmonella typhi, causing general weakness, high fever, a rash of red spots on the chest and abdomen |
| *Ulcers* | a break in the skin extending to all its layers, or a break in the mucous membrane lining the alimentary tract, that fails to heal and is often accompained by inflammation |
| *Urethrorrhoea* | abnormal discharges from the urethra |
| *Urticaria* | an itchy rash resulting from the release of histamine by mast cells |
| *Uteralgia* | pain in the uterus |
| *Veneral diseases* | any disease transmitted by sexual intercourse |
| *Verminosis* | infestation with vermin |
| *Vertigo* | dizziness |
| *Vesical calculi* | stone in the urinary bladder |
| *Vomiting* | the reflex action of ejecting the contents of the stomach through the mouth |
| *Vulnerary* | an agent that promotes healing of wounds |
| *Warts* | a circumscribed cutaneous excrescence |
| *Whooping cough* | an acute infectious disease, primarily of infants, caused by the bacterium Bordetella pertusis, which affects the respiratory tract |
| *Wounds* | a break in the structure of an organ or tissue caused by an external agent |

# BIBLIOGRAPHY


Abraham, Z. 1981. Ethnobotany of the Todas, the Kotas and the lrulas of the Nilgiris, 308 – 320. In S.K. Jain (Ed.) *Glimpses of Indian Ethnobotany.*

Aditya, P., A. Kumar & D.K. Yadav. 2006. Herbal renedies for Dental care system in Gaya district, Bihar. *J. Econ. Taxon. Bot.* 30(Suppl.): 167-170.

Agarwal, V.S. & Arya, M. 2005. *Alternanthera sessilis* DC. (Amaranthaceae): A new source for the treatment of rickets and marasmus. *Ethnobotany.* 17:200-201.

Ahmad, A. 1995. Drugs of plant origin as used by certain tribes of eastern (Purvanchal) U.P. (Part -IV). *Int J Mendel* 12 (1-4): 12 – 14.

Ahmad, A. 1996. Drugs of plant origin as used by certain tribes of eastern (Purvanchal) U. P. (Part -V). *Indian J Appl Pure Biol:* 11: 41-52.

Aminduddin, Girach, R. D. & Khan, 1993. Treatment of Malaria through herbal drugs from Orissa. *Fitoterapia* 64: 545 – 548.

Apparanantham, T. & V. Chelladurai, 1986. Glimpses on folk medicines of Dharmapuri forest division, Tamil Nadu, *Anc. Sci. Life.* 5: 182 – 185.

Arinathan, V., A. Amuthavalli & M. Malaiarasi. 2006. Studies of medicinal plants in Thoothukudi, Tamilnadu. *J. Econ. Taxon. Bot.* 30(Suppl.): 353-357.

Arinathan, V., J. Joseph Visuvasam & R. G. Muthu. 2006. Survey of medicinal plants in Kulathur, Tamilnadu. *J. Econ. Taxon. Bot.* 30(Suppl.): 370-374.

Audichya, K.C., K.V. Billore, T.H. Joseph & D.D. Chaturvedi, 1983. Role of indigenous folk remedies for certain acute illnesses in primary health care. *Nagarjun.* 26: 199-201.

Baruah, P. & G.C. Sarma. 1984. Studies on the medicinal uses of plants by the Boro tribals of Assam II. *J. Econ. Taxon. Bot.* 5: 599 – 604.

Baruah, Parukutty & G. C. Sarma, 1987. Studies on the medicinal uses of plants by the Northeast tribes – III. *J. Econ. Taxon. Bot.* 11: 71– 76.

Baruah, M.K., P.D. Choudhury & G.C. Sarma 2006. Ethnomedicinal plants used by Khasi tribe of Cachar district, Assam. *J. Econ. Taxon. Bot.* 30(Suppl.): 110-114.

Basak, S. K. 1997. Medicinal plants of Bankura (W. B.) and their uses. *J. Nat Bot Soc.* 51: 61 – 68.

Basak, S.K. 2006. Medicinal plants of Bankura (West Bengal) and their uses. *J. Econ. Taxon. Bot.* 30(Suppl.): 217-224.

Bedi, S.J. 1978. Ethnobotany of the Ratan Mahal Hills, Gujarat, India, *Econ. Bot.* 32: 278 – 284.

Bennet, S.S.R. 1985. Ethnobotanical studies in West Sikkim. *J. Econ. Taxon. Bot.*7: 317 – 321.

Bhalla, N.P., T.R. Sahu, G.P. Mishra & R.N. Dakwale. 1982. Traditional Plant medicines of Sagar Distt. Madhya Pradesh, India. *J. Econ. Taxon. Bot.* 3: 23 – 32.

Bhatt, R.P. & S.D. Sabnis, 1987. Contribution to the Ethnobotany of Khedbrahma region of north Gujarat. *J. Econ. Taxon. Bot.* 9: 139 – 145.

Bhattacharjee, S., K.C. Tiwari, R. Majuumdar & A.K. Mishra. 1980. Folklore medicine from district Kamrup (Assam). *Bull. Medico – ethnobot. Res.* 1: 447 – 460.

Billore, K.V. & K.C. Audichya. 1978. Some oral contraceptives; family planning tribal way, *J. Res. Indian Med. Yoga & Homeo.* 13: 104.

Billore, K.V., Joseph, T.G. & Dave, S.K. 1998. Interesting folk remedies by the 'Lok, Vaidyas' of Rajasthan for 'Swas Roga'. *Ethnobotany.* 10: 42 – 45.

Borthakur, S.K. 1976. Less known medicinal uses of plants among the tribals of Mikir Hills. *Bull. Bot. Surv. India.* 18: 166 -171.

Borthakur, S.K. Nath, K. & Gogoi, P. 1996. Herbal remedies of the Nepalese of Assam, *fitoerapia.*67: 231 – 237.

Caius, J.F. 1998. The Medicinal and Poisonous Plants of India, Scientific Publishers (India), Jodhpur.

Chaghtai, S.A. & Javed Ahmad. 1978. An ethnobotanical study of Baghrendah (*Jatropha curcas* L.) *J. Res. Indian Med. Yoga & Homeo.* 13: 134 – 136.

Chandra, K. 1995. An Ethnobotanical study on some medicinal plants of District Palamau (Bihar). *Sachitra Ayurved.* 48: 311 – 314.

Chatterjee, S. & A. Bhatterjee. 2006. Some Ethnomedicinal plants of North 24 Paragana, West Bengal. *J. Econ. Taxon. Bot.* 30(Suppl.): 281-282.

Chaudhuri, Rai H.N., D.C. Pal & Tarafder. 1975. Less known uses of plants from the tribal areas of Orissa. *Bull. Bot. Surv India.* 17: 132 – 136.

Chaudhuri, Rai H.N. & D.C. Pal. 1976. Preliminary observations on Ethnobotany of Midnapur district West Bengal. *India. Mus. Bull.* 11: 51 – 53.

Chaudhuri, Rai H.N., H.A. Molla, D.C. Pal & B. Roy. 1982. Plants used in traditional medicine by some tribals of Jalpaiguri district, West Bengal. *Bull. Bot. Surv. India.* 24: 87 – 90.

Chaudhuri, Rai, H.N., D.C. Pal, N.C. Saha & B. Roy. 1985. Some wild edible plants in Calcutta markets. *Man & Life.* 11: 45 – 58.

Chelvan, P.T. 1998. Traditional Phytotherapy among the migrant Tamilian settlers in South Gujarat. *Biojournal* 10: 9-14.

Chhetri, D.R. 2005. Ethnomedicinal plants of the Khangchendzonga National Park, *Sikkim, India. Ethnobotany.* 17: 96-103.

Chopra, R.N. Nayar, S.L. and Chopra, I.C. 1956. Glossary of Indian medicinal plants. CSIR.

Chopra, R.N. Badhwar, R.L. and Ghosh, S. 1965. Poisonous plants of India. Vol. I & II. ICAR, New Delhi.

Dar, G.H., Vir jee, P. Kachroo & G.M. Buth. 1984. Ethnobotany of Kashmir – I. Sind Valley. *J. Econ. Taxon. Bot.* 5: 668 – 675.

Das, S.N., K.P. Janardhanan & S.C. Roy. 1983. Some observations on the ethnobotany of the tribes of Totopara and adjoining areas in Jalpaiguri district, West Bengal. *J. Econ. Taxon. Bot.* 4: 453 – 474.

Das, S.R. 1995. Some beneficial weeds of cultivation from Nadia district, West Bengal. *Bull. Med. Ethnobot. Res.* 16: 74 – 81.

Das, A.K., B.K. Dutta & G.D. Sharma. 2006. Study of Medicinal plants used by the different tribes/ communities of Cachar district Assam. *J. Econ. Taxon. Bot.* 30 (Suppl.) : 294-309.

Deka, D., G.C. Sarma, N. Devi & N. Pathak. 2006. Indigenous herbal medicines used against Malaria in Goalpara and Morigaon district of Assam. *J. Econ. Taxon. Bot.* 30(Suppl.): 177-183.

Deokule, S.S. 2006. Ethno-medicinal plants of Baramati region of Pune district, Maharashtra. *J. Econ. Taxon. Bot.* 30(Suppl.): 59-69.

Duthie, J.F. 1994 (Reprint). Flora of the Upper Gangetic plain. Bishen Singh Mahendra Pal Singh, Dehradun.

Dutta Choudhury, M. & S. Shil. 2006. Some traditional antihepatitis plants used by the People of Manipur. *J. Econ. Taxon. Bot.* 30(Suppl.): 289-293.

Dwivedi, S. N., Dwivedi, S. and Patel, P.C. 2005. Herbal remedies used by the rural people in the treatment of joint diseases. *Ethnobotany.* 17: 193-196.

E. Vargese SVD. 1996. Applied ethnobotany-A case study among the Kharias of central India. *Deep publications.* New Delhi.

Ganesan S., Venkateshan G & Banumathy N. 2006. Medicinal plants used by ethnic group Thottianaickans of Semmalai hills (reserved forest), Tirucirappalli district, Tamil Nadu. *IJTK.* 5(2): 245-252.

Gaur, R.D., M.P. Sharma & J.K. Semwal 1980. Ethnotoxic plants of Garhwal hills in India. *East. Anthrop.* 33: 159 – 163.

Ghosh, A. 2006. Medicinal plants used for treatment of Diabetes by the tribals of Bankura, Purulia and Medinipur of West Bengal. *J. Econ. Taxon. Bot.* 30(Suppl.): 233-238.

Goel, A.K., A.K. Sahoo & V. Mudgal 1984. A contribution to the ethnobotany of Santal Pargana, Bihar. *Bot. Surv. India.*

Goel, A.K. & V. Mudgal 1988. A survey of medicinal plants used by the tribals of Santal Pargana (Bihar). *J. Econ. Taxon. Bot.* 12: 329 – 335.

Gogoi, P. & C.L. Boissya 1984. Informations about a few herbal medicines used by the people of Assam (India) against Jaundice, *Himal. Res. Dev.* 2: 41 – 44.

Gogte, V.V.M. Vaidya 2000. Ayurvedic pharmacology and Therapeutic uses of medicinal plants (Dravyaguna Vignyan). Bhartiya Vidya Bhawan, Mumbai.

Gopakumar, K. ,Vijayalakshmi, B., Shantha, T. R. & Yoganarasimhan, S.N. 1991. Plants used in Ayurveda from Chikmangalur district, Karnataka – II. *J. Econ. Taxon. Bot.* 15: 379 – 389.

Gopal, G.V. & G.L. Shah. 1985. Some folk medicinal plants used for jaundice in Gujarat, India. *J. Res. Edu. Indian Med.* 4: 45 – 49.

Gupta, S.P. 1981. Native medicinal uses of plants by the Asurs of Netarhat plateau (Bihar), 218-231. In S.K. Jain (Ed.) *Glimpses of Indian Ethnobotany.*

Gupta, Rajendra. 1981. Plants in folk medicine of the Himalayas. 83-90. In S.K. Jain (Ed.) *Glimpses of Indian Ethnobotany.*

Gupta, V.C., Hussain, S.J. & Imam, S. 1997. Medico – ethonbotanical survey at Paderu forests of Araku Valley, Andhra Pradesh, *Fitoterapia.* 68: 45 – 48.

Gupta, A.K., Mishra, S.K. & Khan, A.A. 1999. Ethnobotanical notes on some herbs from Chhatisgarh region of Madhya Pradesh. *Adv Plant Sci.* 12: 163-166.

Gupta, A.K. & Tandon, N. 2004 (Ed.). Reviews on Indian Medicinal Plants. Vol. I, II, III & IV. *Indian Council of Medical Research, New Delhi.*

Gupta, A.K. 2005 (Ed.). Quality standards of Indian Medicinal Plants. *Vol.* I, II & III . *Indian Council of Medical Research. New Delhi.*

Gupta, V. 2006. Traditional medicinal plants of the Bangnis of East Kameng district, Arunachal Pradesh. *J. Econ. Taxon. Bot.* 30(Suppl.):310-319.

Handa, S.S. & Kapoor, V.K. 1999. Pharmacognosy. Vallabh Prakashan, Delhi.

Harja, P.K. & A.K. Baishya 1981. Ethnobotanical notes on the Miris (Mishings) of Assam Plants, 161-169. In S.K. Jain (Ed.) *Glimpses of Inidian Ethnobotany.*

Hemadri, K. & S.S. Rao 1983. Leucorrhoea and menorrhagia : Tribal medicine. *Anc. Sci. Life.* 3: 40 – 41.

Hemadri, K. & S.S. Rao 1984. Jaundice: Tribal medicine. *Anc. Sci. Life.*3: 209– 212.

Hemadri, K. 1992. Tribals of Andhra Pradesh: their knowledge in nutritional and medicinal herbs. *Indian Med.* 4(1):1– 6.

Hembrom, P.P. & Goel, A.K. 2005. Horopathy : Ethnomedicine of Mundas. *Ethnobotany* 17: 89 – 95.

Jadeja, B.A., N.K. Odedra & S.I. Choudhary 2006a. Ethnomedicinal plants used by the tribals of Gujarat to cure digestive disorders. *J. Econ. Taxon. Bot.* 30(Suppl.): 115-123.

Jadeja, B. A., N. K. Odedra, H. N. Mori & D. C. Bhati 2006b. Medicinal plants of Gujarat used in Gout. *J. Econ. Taxon. Bot.* 30(Suppl.): 225-232.

Jadhav, D. 2006 a. Ethnomedicinal plants used by Bhil tribe of Bibdod, Madhya pradesh. *IJTK.* 5(2): 263 – 267.

Jadhav, D. 2006 b. Unreported ethnomedicinal uses of *Calotropis gigantea* (L.) R. Br. (Asclepiadaceae) among the tribals of Ratlam district, Madhya pradesh. *Jou.of NTFPs.* 13(1): 53 – 54.

Jadhav, D. 2006 c. Plant sources used for the treatment of types of fevers by Bhil tribe of Ratlam district, Madhya pradesh. *J. Econ. Taxon. Bot.* 30(4): 909–911.

Jadhav, D. 2006 d. Ethnomedicinal plants used for curing skin afflictions by Bhil tribe of Ratlam district, Madhya pradesh. *J. Econ. Taxon. Bot.* 30(Supp.).

Jain, S.K. 1963 a. Studies in Indian ethnobotany – Less known uses of 50 common plants from tribal areas of Madhya Pradesh. *Bull. Bot. Surv. India.* 5: 223 – 226.

Jain, S.K. 1963 b. Studies in Indian ethnobotany – II. Plants used in medicine by the tribals of Madhya Pradesh. *Bull. Reg. Res. Lab. Jammu.* 1: 126 128.

Jain, S.K. 1965 a. Medicinal plantlore of the tribals of Bastar. *Econ. Bot.*19: 236 -250.

Jain, S.K. 1965 b. On the prospects of some new or less known medicinal plant resources. *Indian Med. J.* 59: 270 – 272.

Jain, S.K. & J.N. De 1966. Observations on Ethnobotany of Purulia district, West Bengal. *Bull. Bot. Suv. India.* 8: 237 – 251.

Jain, S.K. & C.R. Tarafder 1970. Medicinal plant lore of the Santals. A revival of P O Bodding's work. *Econ. Bot.* 24: 241 – 278.

Jain, S.K., D.K. Banerjee & D.C. Pal 1973. Medicinal plants among certain Adibasis in India. *Bull. Bot. Suv. India.* 15: 85 – 91.

Jain, S.K. & N. Dam 1979. Some ethnobotanical notes from north – eastern India. *Econ. Bot.* 33: 52 – 56.

Jain, S.K. & S.K. Borthakur 1980. Ethnobotany of the Mikirs of India. *Econ. Bot.* 34: 264 – 272.

Jain, S.K. 1981 (Ed.). Glimpses of Indian ethnobotany. Oxford and IBH Publishing Co., New Delhi

Jain, S.K. 1984. Ethnobotany of Morni and Kalesar (Ambala – Haryana). *J. Econ. Taxon. Bot.* 5: 809 – 813.

Jain, S.K. 1991. Dictionary of Indian folk Medicine and ethnobotany. Deep Publication, New Delhi.

Jain, S.K. 2004 (Ed.). A Manual of Ethnobotany. Scientific Publishers, Jodhpur.

Jain, P. & Sahu, T.R. 1993. An ethnobotanical study of Noradehi Sanctuary Park of M.P., India: Native plant remedies for scorpion sting and snake bite. *J. Econ. Taxon. Bot.* 17: 315 – 328.

Janardhanan, K.P. 1963. An enumeration of the medicinal plants of Khed Taluka, Maharashtra State. *Bull. Bot. Surv. India.*5: 363 – 374.

Jha, P.K. Choudhary, R.S. & Choudhary, S.K. 1997. Studies of Medicinal plants of Palamau (Bihar)- (Part – II). *Bijournal.* 9: 21 – 38.

Joshi, M.C., M.B. Patel & P.J. Mehta 1980. Some folk medicines of Dangs, Gujarat State. *Bull. Medico – ethnobot. Res.* 1: 8 – 24.

Joshi, P. 1982. An ethnobotanical study of Bhils– A preliminary survey. *J. Econ. Taxon. Bot.* 3: 257 – 266.

Jubilee, P. & Subhan, C.N. 2006. Biological activities of ethnomedicinal claims of some plant species of Assam. *IJTK.* 5(2):229-236.

Kadavul, K. & P. Jisha 2006. Medicinal plants and their conservation in Mahe region, U. T. of Pondicherry, Malabar Coast, South India. *J. Econ. Taxon. Bot.* 30(Suppl.): 203-207.

Kataria, R.L. & D.P. Gauchan 2006. Ethnomedicinal plants used in Dysentery and diarrhoea in Hanumangarh district, Rajasthan. *J. Econ. Taxon. Bot.* 30(Suppl.): 128-132.

Kamble, S.Y. & S.G. Pradhan 1980. Ethnobotany of Korkus in Maharashtra. *Bull. Bot. Surv. India.* 22: 201 – 202.

Kapur, S.K. 1993. Ethno – Medico plants of Kangra Valley (Himachal Pradesh). *J. Econ. Taxon. Bot.*17: 395 – 408.

Karnick, C.R., K.C. Tiwari, R. Majumdar & S. Bhattacharjee 1981. Newer ethnobotanical and folklore studies of some medicinal plants of Gauhati and surrounding areas. *Nagarjun.* 24: 240 – 245.

Katewa, S.S. & Arora, A. 1997. Some plants in folk medicines of Udaipur district. (Rajasthan). *Ethnobotany.* 9: 48 – 51.

Katewa, S.S. & Sharma, R. 1998. Ethnomedicinal observations from certain Watershed areas of Rajasthan. *Ethnobotany.*10: 46 – 49.

Kharkongor, P. & J. Joseph 1981. Folklore medico – botany of rural Khasi and Jaintia tribes in Meghalaya. 124 – 136. In S.K. Jain (Ed.) *Glimpses of Indian Ethnobotany.*

Khory, J.K. 1981. Materia Medica of India & their Therapeutics, Neeraj Publishing House, New Delhi.

Kirtikar, K.R. & Basu, B.D. 2004. Indian medicinal plants, Vol. I, II, III &IV (Reprint). Bishen Singh Mahendra Pal Singh, Dehradun.

Krishna, B. & S. Singh 1987. Ethnobotanical observations in Sikkim. *J. Econ. Taxon. Bot.*9: 1 – 7.

Kumar, Y., K. Haridasan & R.R. Rao 1980. Ethnobotanical notes on certain medicinal plants among some Garo people around Balphakram Sanctury in Meghalaya. *Bull. Bot. Surv. India.*22: 161 – 165.

Kumar, Y., S. Fancy & R.R. Rao 1987. Further contribution to the ethnobotany of Meghalaya: Plants used by War Jaintia of Jaintia Hill district. *J. Econ. Taxon. Bot.*11: 65 – 70.

Kumar, M. 1992. Ethnomedicinal studies of plants of chotanagpur plateau, Bihar, *Int. J. Mendel. 9: 137 – 139.*

Kumar, V. & Jain, S.K. 1998. A contribution to ethnobotany of Surguja district in M.P. *Ethnobotany.*10: 89 – 96.

Kumar, S. & Chauhan, A.K.K. 2005. Medicinal plants used by local inhabitants in Bharatpur district, Rajasthan. *Ethnobo.*17: 179 – 183.

Kumar, A., A. Singh, Y. Sharma & J.C. Rana 2006. Ethno-medico-botany of some important plants in Mandhala watershed of Himachal Pradesh. *J. Econ. Taxon. Bot.* 30(Suppl.): 145-150.

Lakshamanan, K.K. & A.S. Sankaranarayanan 1988. Some folklore medicines in the remote hemlets, Dhoomanoor & Champukarai of Anai Katty hills, Coimbatore, Tamil Nadu. *Indian J. Forestry.* 11: 217 – 219.

Lal, S.D. & B.K. Yadav 1983. Folk medicines of Kurukshetra District (Haryana), India. *Econ. Bot.* 37: 299 – 305.

Lal, S.D. & K. Lata 1980. Plants used by the Bhat Community for regulating fertility. *Econ. Bot.*34: 273 – 275.

Lindley, John 1981. Flora Medica, Ajay Book Service, New Delhi.

Maheshwari, J. K., K. K. Singh & S. Saha 1980. Ethnomedicinal uses of plants by Tharus in Kheri district, U. P. *Bull. Medico-ethnobot. Res.* 1:318-337.

Maheshwari, J.K., K.K. Singh & S. Saha 1981. The Ethonbotany of the Tharus of Kheri district. Uttar Pradesh. *Nation. Bot. Res. Inst.* Lucknow.

Maheshwari, J.K. & J.P. Singh 1984. Contribution to the ethnobotany of Bhoxa tribe of Bijnor & Pauri Garhwal district, U.P. *J. Econ. Taxon. Bot.* 5: 251 – 259.

Maheshwari, J.K. B.S. Kalakoti & Brijlal 1986. Ethnomedicine of Bhil Tribe of Jhabua district, Madhya Pradesh. *Anc. Sci. Life.* 5: 255 – 261.

Maheshwari, J.K. K.K. Singh & S. Saha. 1986. Ethnobotany of tribals of Mirzapur district. U.P. *Nation. Bot. Res. Inst.,* Lucknow.

Maheshwari, J.K. & J.P. Singh 1987. Traditional phytotherapy amongst the Kol tribe of Banda district, Uttar Pradesh. *J. Econ. Taxon. Bot.* 9: 165 – 171.

Malhotra, S.K. & S. Moorthy 1973. Some useful and medicinal plants of Chandrapur district (Maharasthra State). *Bull. Bot. Surv. India.*15: 13 – 21.

Megoneitso & R.R. Rao. 1983. Ethnobotanical studies in Nagaland – 4. Sixty two medicinal plants used by the Angami Nagas. *J. Econ. Taxon. Bot.* 4: 167 – 172.

Mishra, R. & R.D. Dixit 1973. Vajradanti – An ethnobotanical study. I. *J. Res. Indian Med.* 8: 89 – 91.

Mishra, R. & K.V. Billore 1983. Some ethnobotanical lores from Banswara district. Rajasthan. *Nagarjun.* 26: 229 – 231.

Mishra, D.P. & T.R. Sahu 1984. Euphorbiaceous plants used in medicine by the tribals of Madhya Pradesh, India. *J. Econ. Taxon. Bot.* 5: 791 – 794.

Mishra, S.K. & Verma, R.M.P. 1995. Medicinal plants of Champaran (Bihar) used in various diseases. *Flora fauna.*1: 75 – 76.

Mishra, B. & Vaisya, R. 2004. Bhavaprakasa, Vol. I & II. Chaukhambha Sanskrit Sansthan, Varanasi.

Mohanty, R.B. Padhy, S.N. & Dash, S.K. 1996. Traditional phytotherapy for diarrhoeal diseases in Ganjan and Phulbani district of South Orissa, India. *Ethnobotany.*8: 60 – 65.

Mokat, D. N. & S. S. Deokule 2006. Ethno-medico-botanical survey of Ratnagiri district of Maharashtra State. *J. Econ. Taxon. Bot.* 30(Suppl.): 140-144.

Molla, H.A. & B. Roy 1985. Traditional uses of some medicinal plants by the Rabha tribals in Jalpaiguri district, West Bengal. *J. Econ. Taxon. Bot.*7: 578 – 580.

Mudgal, V. & D.C. Pal 1980. Medicinal plants used by tribals of Mayurbhanj (Orissa). *Bull. Bot. Surv. India.* 22: 59 – 62.

Mudgal, V. Khanna, K.K. & Hajra, P.K. 1997. Flora of Madhya Pradesh. Vol. II. BSI. Calcutta.

Mukerji, B. 1953. The Indian Pharmaceutical Codex, New Delhi.

Mukherjee, A.S. Mukherjee & M. Mukherjee 1986. Less known medicinal plants of Rubiaceae in Darjeeling Hills, West Bengal (India). *Indian J. Appl. & Pure. Bot.*1: 86 – 91.

Murthy, K.S., P.C. Sharam & Prem Kishore 1986. Tribal remedies for sanke bite from Orissa. *Anc. Sci. Life.* 6: 122 – 123.

Nagar, M. 1985. The use of wild plant foods by aboriginal communities in Central India, 337 – 342. In Mishra, V.M. & Belwood, P. (Ed.) *Recent Adv. In Indo-Pacific Prehist.* Oxford & IBH, New Delhi.

Nagarjun, N. & Rao, K.N. 1989. Folk - medicine for diabetes from Rayalaseema of Andhra Pradesh, *Ancient. Sci. Life.* 9: 31 – 35.

Nagendra Prasad, P. & Z. Abraham 1984. Ethnobotany of Nayadis of North Kerala. *J. Econ. Taxon. Bot.* 5: 41 – 48.

Namjoshi, A.N. 1960. Studies in Pharmacognosy of Ayurvedic Drugs. Board of Research on Ayurveda, Bombay.

Narain, S. & Singh, U. 2006a. Contribution to the Ethnobotanical plants of Mirzapur district (U.P.). *J. Econ. Taxon. Bot.* 30(Suppl.): 14 – 17.

Narain, S. & Singh, U. 2006b. Contribution to the Ethnobotanical plants of Sonbhadra district (U.P.). *J. Econ. Taxon. Bot.* 30(Suppl.): 18 – 20.

Nath, P. C., Sarmah, R., Arunachalam, A. & Barbhuiya, A. 2006. NTFPs: Potential and their conservation in Dihing-patkai wildlife santuary of Assam. *Jou. of NTFPs.* 13: 17-23.

Nautiyal, S. 1981. Some medicinal plants of Garhwal hills – A traditional use. *J. Sci. Res. Pl. Med.*2: 12 – 18.

Negi, K.S., J.K. Tiwari & R.D. Gaur 1985. Economic importance of some common trees in Garhwal, Himalaya: An ethnobotanical study. *Indian J. Forestry.* 8: 276 – 289.

Oommachan, M. 1977. The Flora of Bhopal. J.K. Jain Brothers, Bhopal.

Pal, D.C. & J.N. Srivastava. 1976. Preliminary notes of ethnobotany of Singhbhum district, Bihar, *Bull. Bot. Surv. India.*18. 247 – 250.

Pal, D.C. 1980. Observations on folklore about plants used in veterinary medicine in Bengal, Orissa & Bihar. *Bull. Bot. Surv. India.* 22: 96 – 99.

Pal, G.D. 1984. Observations on ethnobotany of tribals of Subansiri. Arunachal Pradesh. *Bull. Bot. Surv. India.*26: 26 - 37.

Pal, D.C. & S.K. Jain 1989. Notes on Lodha medicine in Midnapur district. W.B. India. *Econ. Bot.* 43: 464 – 470.

Pandey, G.S. 1969. Bhavaprakash Nighantu (With Commentary by Dr. K.C. Chunekar). Chowkhamba Sanskrit Sansthan, Varanasi.

Pandit, B.R. Kotiwar, O.S., Oza, R. A. & Kumar, R.M. 1996. Ethno-medicinal plant lore from Gir forest, Gujarat. *Adv plant sci.* 9: 81 – 84.

Panigrahi, A.K. & Alka, S. 2000. Glossary of useful and economically important plants. New Central Book agency. Calcutta.

Paranjpe, P. 2005. Indian medicinal plants: forgotten healers. Chaukhamba Sanskrit Pratisthan, Delhi.

Pawar, S. & D.A. Patil 2006. Ethnobotanical aspects of naturalised exotic plants of Jalgaon district, Maharashtra. *Jou. Of NTFPs.* 13:55-59.

Pearsall, J. & Trumble, B. 1996. The oxford english Reference Dictionary. Oxford University Press, Oxford.

Pereira, N. 2006. Ethnobotanical relevance in tribal life: A study on Warli tribe of Thane district, Maharashtra. *J. Econ. Taxon. Bot.* 30(Suppl.): 31-39.

Peter Paul Hembrom & A.K. Goel 2005. Horopathy: Ethnomedicine of Mundas. *Ethnobotany*.17:89-95.

Prakash, A. & K.K. Singh 2006. Useful medicinal plants for general tonic, sexual vigour, strength, vitality and venereal diseases among the tribals of Uttar Pradesh and Uttaranchal. *J. Econ. Taxon. Bot.* 30(Suppl.): 326-331.

Printya, J.V. 1963. The Sushruta Samhita. Vol. I & II, Motilal Banarsidas. Delhi.

Puri, H.S. 1983. Medicinal plants of Tejpur (Assam). *Bull. Medico-ethnobot. Res.*4: 1–13.

Purkayastha, J. & Subhan, C.N. 2005. Biological activitis of ethnomedicinal claims of some plant species of Assam. *IJTK.* 5(2), 229 – 236.

Purohit, V.P., R.A. Silas & R.D. Gaur 1985. Ethnobotanical studies of some medicinal plants used in skin diseases from Raath (Pauri) Garhwal, Himalaya. *J. Sci. Res. Pl. Med.* 6: 39 – 47.

Ramachandran, V.S. & N.C. Nair 1981 a. Ethnobotanical studies in cannanore district, Kerala state (India). *J. Econ. Taxon. Bot.* 2: 65 - 72.

Ramachandran, V.S. & N.C. Nair 1981 b. Ethnobotanical observations on 'lrulars' of Tamil Nadu (India). *J. Econ. Taxon. Bot.*2: 183 – 190.

Rana, T.S., Datt. & R.R. Rao 1996. Strategies for sustainable utilization of plant resources by the tribals of the Tons – Valley, Western Himalayas. *Ethnobotany.* 8: 96 – 104.

Rai, M.K. 1985. Plants used as medicine by tribals of Chhindwara District (M.P.). *J. Econ. Taxon. Bot.* 7: 385 – 387.

Rajendran, K. & S. K. Rengamani. 2006. Medicinal plants and their utilization by villagers in Southern districts of Tamilnadu. *J. Econ. Taxon. Bot.* 30(Suppl.): 208-216.

Rajwar, G.S. 1983. Low altitude medicinal plants of South Garhwal. *Bull. Medico-ethnobot. Res.*4: 14 – 28.

Rao, R.R. & B. Neogi, 1980. Observations on the ethnobotany of the Khasi & Garo tribes in Meghalaya. *J. Econ. Taxon. Bot.*1: 157 – 162.

Rao, R.R. 1981 a. Ethnobotanical stuidies on the flora of Meghalaya – Some interesting reports of herbal medicines. In S.K. Jain (Ed.) *Glimpses of Indian Ethnobotany.* pp. 137-148.

Rao, R.R. 1981b. Ethnobotany of Meghalaya – Medicinal plants used by Khasi and Garo tribes. *Econ. Bot.* 35: 4 – 9.

Rao, R.R. & N.S. Jamir 1982a. Ethnobotanical studies in Nagaland – II: 54 Medicinal plants used by Nagas. *J. Econ. Taxon. Bot.* 3: 11 – 17.

Rao, R.R. & N.S. Jamir 1982b. Ethnobotanical studies in Nagaland – I: Medicinal plants. *J. Econ. Bot.* 36: 176 – 181.

Rastogi & Mehrotra 1997. Compendium of Indian medicine Plant, Vol. I, II, III, IV & V. CDRI, Lucknow and PID Delhi.

Roy, G.P. & K.K. Chuturvedi 1986a. Plants used as tonic by the tribals of Abujh Marh Reserve area, Bastar district in Madhya Pradesh. *Vanyajati.* 34: 27-32.

Roy, G.P. & K.K. Chaturvedi 1986b. Ethnomedicinal trees of Abujh-Marh area, Madhya Pradesh. *Folklore.* 27: 95 – 100.

Roy, G.P. & K.K. Chaturvedi 1987. Less known medicinal uses of rare and endangered plants of Abujh Marh reserve area, Bastar (Madhya Pradesh). *J. Econ. Taxon. Bot.* 9: 325 – 328.

Samvatsar, Swati 1996. The Flora of Western tribal Madhya Pradesh, Scientific Publishers (India), Jodhpur.

Sarma, S.K., B. Bevi & D.K. Bhattacharya 2006. Ethnomedicinal uses of plants by the Sarania tribe of Nalbari district of Assam. *J. Econ. Taxon. Bot.* 30(Suppl.): 133-139.

Saxena, H.O. & P.K. Dutta 1975. Studies on the ethnobotany of Orissa. *Bull. Bot. Surv. India.* 17: 124 – 131.

Saxena, A.P. & K.M. Vyas 1981. Ethnobotanical records on infections diseases from tribals of Banda district. *J. Econ. Taxon. Bot.* 2: 191 – 194.

Saxena, H.O., M. Brahman & P.K. Dutta 1931. Ethnobotanical studies in Orissa, In S.K. Jain (Ed.). *Glimpses of Indian Ethnobotany.* pp. 232 – 244.

Saxena, A.P. & K.M. Vyas 1983. Ethnobotany of Dhasan Valley. *J. Econ. Taxon. Bot.* 4: 121 – 128.

Saxena, A.P. & K.M. Vyas 1986. Antimicrobial activity of seeds of some ethno-medicinal plants. *J. Econ. Taxon. Bot.* 8: 291 – 299.

Saxena, H.O. 1986. Observations on the ethnobotany of Madhya Pradesh. *Bull. Bot. Surv. India.* 28: 149 – 156.

Sebastian, M.K. & M.M. Bhandari 1984. Medico-ethnobotany of Mount Abu, Rajasthan, India. *J. Ethnopharmacol.* 12: 223 – 230.

Sebastian, M.K. 1984. Plants used as veterinary medicines, galactagogues and fodder in the forest areas of Rajasthan. *J. Econ. Taxon. Bot.* 5: 785 – 788.

Sen, R., D. C. Pal. & B. Roy 1984. Botany and ethnobotany of Parashi. *J. Econ. Taxon. Bot.* 5 : 857 – 858.

Sen, R.S. Bandopadhayay, E. Biswas & A. Guha 1985. Ethnobotanical uses of Herbaria – IV. *J. Econ. Taxon. Bot.* 6: 331 – 335.

Shah, N.C. & M.C. Joshi 1971. An ethnobotanical study of Kumaon region of India. *Econ. Bot.* 25: 414 – 422.

Shah, G.L., A.R. Menon & G.V. Gopal 1981. An account of the ethnobotany of Saurashtra in Gujarat state (India). *J. Econ. Taxon. Bot.* 2: 173 – 182.

Shah, N.C. 1982. Herbal folk medicines in northern India. *J. Ethnopharmacol.* 6: 293 – 301.

Shah, G.L., S.S. Yadav & N. Badri 1983. Medicinal plants from Dahanu Forest Division in Maharashtra State. *J. Econ. Taxon. Bot.* 4: 141 – 151.

Shah, G.L. 1984. Some economically important plants of Salsette Island near Bombay. *J. Econ. Taxon. Bot.* 5: 753 – 765.

Shah, G.L. & G.V. Gopal 1986. Folklore medicines of Vasavas, Gujarat, India. *Acta. Bot. Indica.* 14: 48 – 53.

Shah, G.L. & G.V. Gopal 1992. An ethnobotanical profile of the Dangies. *J. Econ. Taxon. Bot.* 3: 355 – 364.

Shah, M.A.U. 2005. Ethnomedicinal plants of Bhakkar Tehsil in Punjab, Pakistan. *Ethnobotany.* 17: 171-175.

Sharma, R.K. & Bhagvan Dash 1976. Charaka samhita, Vol. I & II, Chaukhamba Sanskrit Series, Varansi, India.

Sharma, P.K., S.K. Dhyani & V. Shanker 1979. Some useful and medicinal plants of the district Dehradun and Siwalik. *J. Sci. Res. Pl. Med.* 1: 17 – 43.

Sharma, B.D. & S.K. Malhotra 1984. A contribution to the ethnobotany of tribal areas in Maharashtra. *J. Econ. Taxon. Bot.* 5: 533 – 537.

Sharma, P.V. 2003. Dravyaguna-Vijnana. Chaukhamba Bharti Academy Varanasi.

Sharma, S.C., G.N. Sharma & R.K. Aggarwal 2006. Traditional significant Medicinal plants of Nagaur district (Rajasthan). *J. Econ. Taxon. Bot.* 30 (Suppl.): 320-325.

Shekhawat, G.S. & Sunil Anand 1984. An ethnobotanical profile of Indian Desert. *J. Econ Taxon. Bot.* 5: 591 – 598.

Siddiqui, M.B. & Husian, W. 1991. Traditional treatment of diarrhoea and dysentery through herbal drugs in rural India. *fitoterapia.* 62: 325 – 329.

Singh, Harbhajan 1945. Home useful wild plants of the Delhi Province. *India. J. Agr. Sci.* 15: 297 – 307.

Singh, K.K., S.K. Palvi & H.B. Singh 1980. Survey of some medicinal plants of Dharchula block in Pithoragarh district of U.P. *Bull. Medico-ethnobot. Res.* 1: 1– 7.

Singh, V. & R.P. Pandey 1980. Medicinal plantlore of the tribals of eastern Rajasthan. *J. Econ. Taxon. Bot.* 1: 137 – 147.

Singh, K.K. & J.K. Maheshwari 1983. Traditional phytotherapy amongst the tribals of Varanasi district, Uttar Pradesh. *J. Econ. Taxon. Bot.* 4: 829 – 838.

Singh, K.K. & Saha & J.K. Maheshwari 1985. Ethnobotany of *Helicteres isora* L., in Kheri district, Uttar Pradesh. *J. Econ. Taxon. Bot.* 7: 487 – 492.

Singh, A.K., R.N. Singh & S.K. Singh 1987. Some ethnobotanical plants of Terai region of Gorakhpur, district- I. *J. Econ. Taxon. Bot.* 9: 407 – 410.

Singh, H. 1988. Ethnobiological treatment of Piles by Bhoxas of Uttar Pradesh. *Anc. Sci. Life.* 8: 167 – 170.

Singh, H. & Maheshwari, J.K. 1992. Traditional remedies for snake bite & Scorpion – sting among the Bhoxas of Nainital district. U.P. *Aryavaidyah.* 6: 120 123.

Singh, L.B. Singh, C.L. 1992. An Ethno-Medico-botanical study of Deoghar district (Bihar). *Biojournal.*4: 83 – 86.

Singh, P.B. & Aswal, B.S. 1992. Medicinal plants of Himachal Pradesh used in Pharmaceutical Industry. *Bull. Med. Ethno. Res.*13: 172 – 208.

Singh, P.B. 1993. Medicinal plants of Ayurvedic importance from Mandi district of Himachal Pradesh. *Bull. Med. Ethnobot. Res.*14: 126 – 136.

Singh, K.K. and Prakash, A. 1994. Indigenous phytotherapy among the Gond tribes of Uttar Pradesh, India. *Ethnobotany.*6: 37 – 41.

Singh, K.K. 1997. Studies on native medicine of Jaunsari tribe of Dehradun district, U.P. India. *Int. J. Pharmacog.* 35: 105 – 110.

Singh, H.B.K. 1997. Studies on Medico-botany of Meitel community in Manipur State, India (II). *Adv. Plant. Sci.*10: 13 – 18.

Singh, V.K., Ali, Z.A. & Siddiqui, M.K. 1997. Medicinal plants used by the forest ethnics of Gorakhpur district (U.P.). *Int. J. Pharmacog.*35: 194 – 206.

Singh, S.K., Das. P.K. & Das, M.N. 1998. Ethnobotanical studies on some medicinal plants of Rajgir, Bihar. *Int. J. Mendeol.*15: 145 – 148.

Singh, V. & R.P. Pandey 1998. Ethnobotany of Rajasthan. Scientific Publishers (India), Jodhpur.

Sood, S.K. & S. Thakur 2004. Ethnobotany of Rewalsar Himalaya. Deep Publications. New Delhi.

Srivastava, D.K. & S.K. Varma 1981. An ethnobotanical study of Santhal Pargana, Bihar, *Inidan. Forester.* 107: 30 – 41.

Srivastava, S.C. & Rout, N. 1994. Some plants of ethnopaediatric importance in district Koraput, Orissa. *Bull. Bot. Surv. India.*36: 166 – 168.

Sudhakar, S. & R.S. Rolla 1985. Medicinal plants of Upper East Godavari District, Andhra Pradesh, and need for establishment of Medicinal Farm. *J. Econ. Taxon. Bot.* 7: 399 – 406.

Sudhakar, A. & Chetty, K.M. 1998. Medicinal importance of some angiospermic weeds used by the rural people of Chittoor district of Andhra Pradesh, India. *Fitoterapia.*69: 390 400.

Tarafder, C.R. 1978. Some interesting new uses of the well known plant *Vitex negundo* L. (NISINDA). *Bull. Bot. Surv. India* 20: 176-177.

Tarafder, C.R. & H.N., Rai Chaudhuri 1981. Less known medicinal uses of plants among the tribals of Hazaribagh district of Bihar. 208-217. In S.K. Jain (Ed.). *Glimpses of Indian Ethnobotany.*

Tarafder, C.R. 1985. *Acacia catechu* (L.f.) Willd. (Fabaceae) - Inddigenous and less known uses of the plant. *Folklore.*26: 158 – 159.

Tarafder, C.R. 1983. Traditional medicinal plants used by the tribals of Ranchi & Hazaribagh district, Bihar- Plants used in stomach troubles. *J. Econ. Taxon. Bot.* 4: 891 – 896.

Tarafder, C.R. 1983a. Ethnogynaecology in relation to plants – II. Plants used in abortion. *J. Econ. Taxon. Bot.*4: 507 – 516.

Tarafder, C.R. 1983b. Ethnogynaecology in relation to plants – I. Plants used in antifertility and conception. *J. Econ. Taxon. Bot.* 4: 483 – 490.

Tarafder, C.R. 1986. Ethnobotany of Chhotanagpur (Bihar). *Folklore.* 27: 95-100.

Tarafder, C.R. 1987. Some traditional knowledge about tribal health. *Folklore.* 28: 37-42.

Tarafder, C.R. 1984a. Ethnogynaecology in relation to plants. Part- III. Plants used to accelerate delivery and in pre and post-natal complaints. *J. Econ. Taxon. Bot.* 5: 572 – 576.

Tarafder, C.R. 1984b. Less known nine medicinal plants used by the tribals for curing gonorrhoea in Ranchi and Hazaribagh district, Bihar. *Folklore.*25: 47- 49.

Tarafder, C.R. 1984c. Less known twenty three medicinal plants used by the tribals for curing boils in Ranchi and Hazaribagh district. Bihar. *Vanyajati.* 32: 14-19.

Tiwari, K.C., R. Majumdar & S. Bhattacharjee 1979. Folklore medicine from Assam and Arunachal Pradesh (district Tirap). *Quart. J. Crude Drug Res.*17: 61 – 67.

Tribedi, G.N., R.N. Kayal & H.N. Rai Chaudhuri 1982. Some medicinal plants of Mayurbhanj (Orissa). *Bull. Bot. Surv. India.* 24: 117 – 120.

Tribedi, G.N., H.A. Molla & D.C. Pal 1985. Some uses of plants from the tribal areas of Chotanagpur, Bihar. *Nagarjun.* 29: 15 – 18.

Tripathi *et al.* 1996. Medicinal plants of Rajasthan in Indian system of medicine. *Ancient Sci. Life.*15: 190 – 212.

Udayan, P.S., S. George, K.V. Tushar, S. Sudhakar Raja & I. Balachandran 2006. A note on the Ethnomedicinal plants used by the Kaadar Tribes, Kariyan Shola, Top Slip, Tamil Nadu. *J. Econ. Taxon. Bot.* 30(Suppl.): 5 – 12.

Udayan, P.S., S. George, K.V. Tushar, S. Sudhakar Raja & I. Balachandran 2006b. Ethnobotanical information from the Muduga tribe of Mukkali, near Silent Valley National Park, Palakkad district, Kerala. *J. Econ. Taxon. Bot.* 30(Suppl.):27 –30.

Uniyal, M.R. 1968. Medicinal plants of Bhagirathi valley lying in Uttarkashi forest division. *Indian Forester.* 94: 407 – 420.

Uniyal, M.R. & M.S. Chauhan 1973. Traditionally important medicinal plants of Kangra Valley in Dharamsala forest circle, Himachal Pradesh. *J. Res. Indian. Med.* 8: 76 – 85.

Upadhyay, R. & Jaswant Singh 2005. Ethnomedicinal uses of plants from Tikri forest of Gonda district. *Ethnobotany.*17: 167– 170.

Upadhye, A., M.S. Kumbhojkar & V.D. Vartak 1986. Observations on wild plants used in folk medicine in the rural areas of the Kolhapur district. *Anc. Sci. Life.* 6: 119 – 121.

Upadhay, O.P., Kumar, K. & R.K. Tiwari 1998. Ethnobotanical study of skin treatment uses of medicinal plants of Bihar, *Pharmaceut Biol.* 36: 161 – 172.

Ved Prakash & S.K. Jain 1984. Poaceae: Tribe - Isachneae. *Fasc. Fl. India.* 14: 41 – 42. Bot. Surv. India. Calcutta.

Ved Prakash & B.N. Mehrotra 1987. Ethnobotanical studies on the flora of Khandala, Maharashtra State. *J. Econ. Taxon. Bot.* 9: 205 – 208.

Ved Prakash & B.N. Mehrotra 1988. Unrecorded traditional medicines-I. *Anc. Sci. Life.* 3: 110-112.

Venugopal, K. & K. Raghunathan 1982 (Ed.). Medicobotanical-cum-ethnomedi-cinal survey of district Jhabua (M.P.). *Centr. Counc. Res. Ayurv. Siddha Newsletter.* 4 - 5.

Verma, D.M. Balakrishan, N.P. & R.P. Dixit 1993. Flora of Madhya Pradesh. BSI, Calcatta.

Vir Jee, G.H. Dar, P. Kachroo & G.M. Bhat 1984. Taxo – ethnobotanical studies of the rural areas in district Rajouri (Jammu). *J. Econ. Taxon. Bot.* 5: 831 – 838.

Warrier, P.K., V.P.K. Nambiar & C. Ramankutty 1994. Indian medicinal plants. Vol. I-V. Oriental Longum, Chennai.

Yesodharan, K. & K.A. Sujana 2006. Ethnomedicinal plants used by the trbals of Parambikulam Wildlife Sanctuary, Kerala to cure cuts and wounds. *J. Econ. Taxon. Bot.* 30 (Suppl.): 365-369.

# INDEX OF ENGLISH, HINDI & SANSKRIT NAMES OF PLANTS

| | | | | |
|---|---|---|---|---|
| *Mucuna pruriens* | Cowhage | Kaunc | Kapikacchuh | 168 |
| *Mukia madraspatana* | - | Bilari | Trikosaki | 169 |
| *Murraya koenigii* | Curry leaf tree | Mitha nim | Kaidarya | 170 |
| *Nelumbo nucifera* | Indian lotus | Kamal | Kamala | 171 |
| *Nerium indicum* | Indian oleander | Kaner | Karavira | 173 |
| *Nyctanthes arbor-tristis* | Night jasmine | Harsinghar | Sephali | 174 |
| *Nymphaea nouchali* | Indian red water lily | Koka | Kumuda | 175 |
| *Ocimum basilicum* | Sweet basil | Babui tulsi | Barbari | 176 |
| *Ocimum canum* | Sacred basil | Jungli tulsi | Arjaka | 177 |
| *Ocimum gratissimum* | Shrubby basil | Bantulasi | Vanabarbarika | 178 |
| *Ocimum sanctum* | Holy basil | Tulsi | Tulasi | 179 |
| *Oroxylum indicum* | Indian trumpet tree | Arlu | Syonakah | 181 |
| *Oxalis corniculata* | Indian sorrel | Amrul | Cangeri | 182 |
| *Pandanus fascicularis* | Screw pine | Kevda | Ketaki | 183 |
| *Papaver somniferum* | Opium poppy | Aphim | Aphuka | 184 |
| *Pergularia daemia* | - | Utaran | Kurutakah | 186 |
| *Phyla nodiflora* | Purple lippia | Jalpippali | Jalapippali | 187 |
| *Phyllanthus emblica* | Emblic myrobalan | Amvala | Amalaki | 188 |
| *Phyllanthus niruri* | Country gooseberry | Bhonyabali | Bhudhatri | 189 |
| *Physalis minima* | Country gooseberry | Cirpoti | Cirapotha | 191 |
| *Plumbago zeylanica* | Ceylon leadwort | Chita | Chitrak | 192 |
| *Pongamia pinnata* | Indian beech | Karanj | Karanjah | 193 |
| *Portulaca oleracea* | Common purslane | Badi loni | Brhat loni | 194 |
| *Portulaca quadrifida* | - | Chotaluniya | Laghughonika | 195 |
| *Psidium guajava* | Common guava | Amrud | Perukah | 196 |
| *Psoralea corylifolia* | Babchi seeds | Bakuchi | Bahuchi | 198 |
| *Pterocarpus marsupium* | Indian kino tree | Bijasal | Bijaka | 199 |
| *Rauvolfia serpentina* | Serpentine root | Sarpgandha | Nakuli | 200 |
| *Ricinus communis* | Castor | Erandi | Svetaeranda | 201 |
| *Rubia cordifolia* | Indian madder | Manjit | Manjistha | 202 |
| *Santalum album* | Sandal wood | Safed chandan | Chandanam | 204 |
| *Saraca asoca* | Ashoka tree | Asok | Asoka | 205 |
| *Senna absus* | - | Chaksu | kulatthika | 206 |
| *Senna auriculata* | Tanner's cassia | Tarwar | Avartaki | 207 |
| *Senna occidentalis* | Negro coffee | Kasmudi | Kasamardah | 208 |
| *Senna tora* | Foetid cassia | Chakunda | Cakramardah | 210 |
| *Sesbania grandiflora* | Sesban | Agasti | Agasti | 211 |
| *Sida cordifolia* | Country mallow | Bariara | Bala | 212 |